Principles and Practice of Plant Hormone Analysis

Volume 1

Biological Techniques Series

J. E. TREHERNE
Department of Zoology
University of Cambridge
England

P. H. RUBERY
Department of Biochemistry
University of Cambridge
England

Ion-sensitive Intracellular Microelectrodes, *R. C. Thomas,* 1978
Time-lapse Cinemicroscopy, *P. N. Riddle,* 1979
Immunochemical Methods in the Biological Sciences: Enzymes and Proteins, *R. J. Mayer* and *J. H. Walker,* 1980
Microclimate Measurement for Ecologists, *D. M. Unwin*, 1980
Whole-body Autoradiography, *C. G. Curtis, S. A. M. Cross, R. J. McCulloch* and *G. M. Powell,* 1981
Microelectrode Methods for Intracellular Recording and Ionophoresis, *R. D. Purves,* 1981
Red Cell Membranes—A Methodological Approach, *J. C. Ellory* and *J. D. Young,* 1982
Techniques of Flavonoid Identification, *K. R. Markham*, 1982
Techniques of Calcium Research, *M. V. Thomas,* 1982
Isolation of Membranes and Organelles from Plant Cells, *J. L. Hall* and *A. L. Moore,* 1983
Intracellular Staining of Mammalian Neurones, *A. G. Brown* and *R. E. W. Fyffe,* 1984
Techniques in Photomorphogenesis, *H. Smith* and *M. G. Holmes*, 1984
Principles and Practice of Plant Hormone Analysis, *L. Rivier* and *A. Crozier,* 1987
Wildlife Radio Tagging, *R. Kenward,* 1987
Immunochemical Methods in Cell and Molecular Biology, *R. J. Mayer* and *J. M. Walker,* 1987

Principles and Practice of Plant Hormone Analysis

Volume 1

Edited by

Laurent Rivier
Institut de Biologie et de Physiologie Végétales
Université de Lausanne
Lausanne
Switzerland

Alan Crozier
Department of Botany
The University
Glasgow
UK

1987

ACADEMIC PRESS
Harcourt Brace Jovanovich, Publishers
London Orlando San Diego New York Austin
Boston Sydney Tokyo Toronto

ACADEMIC PRESS INC. (LONDON) LTD.
24–28 Oval Road,
London NW1

United States Edition published by
ACADEMIC PRESS INC.
Orlando, Florida 32887

British Library Cataloguing in Publication Data

Principles and practice of plant hormone analysis. – (Biological techniques series).
1. Plant hormones
I. Rivier, Laurent II. Crozier, Alan
III. Series
581.19′27 QK731

ISBN 0-12-198375-7 v.1
ISBN 0-12-198376-5 v.2
LCCCN 86-072915

Typeset and printed by
W. & G. Baird Ltd., The Greystone Press,
Antrim, Northern Ireland.

Contributors

Alan Crozier, *Department of Botany, The University, Glasgow G12 8QQ, UK*

Peter Hedden[1], *East Malling Research Station, East Malling, Maidstone, Kent ME19 6BJ, UK*

Roger Horgan, *Department of Botany and Microbiology, University College of Wales, Aberystwyth, Dyfed SY23 3DA, UK*

Steven J Neill, *Department of Botany and Microbiology, University College of Wales, Aberystwyth, Dyfed SY23 3DA, UK*

[1] Present address: Long Ashton Research Station, Bristol BS18 9AF, UK

Preface to Volume 1

Details of the types and amounts of hormones present in plant tissues, as well as a knowledge of the biosynthetic and catabolic pathways that control the size of endogenous hormone pools, are key ingredients in an understanding of the processes involved in the regulation of plant growth and development. Unfortunately, obtaining reliable information on these topics is less than straightforward. One of the major obstacles to progress is the formidable task presented by the accurate analysis of trace quantities of hormones in plant extracts which are exceedingly complex multicomponent mixtures. In the circumstances, it is hardly surprising that so many studies on endogenous plant hormones are flawed by gross analytical errors. This has been a central problem in growth regulator research for many years and continues to be the subject of heated debate whenever the physiological significance of endogenous hormones and their levels in plant tissues are discussed.

Recently, there have been rapid advances in the analytical sciences and much of the methodology that has come into widespread use can be applied to the purification and analysis of nanogram and even sub-nanogram amounts of endogenous plant hormones. Concomitant with these practical developments has been an improved understanding of analytical theory in relation to the problems encountered when attempting to identify and measure hormones in plant extracts. As a consequence, plant hormone analysis has progressed from an art to a science and both the practicalities of analytical strategy and the validity of analytical data can now be assessed on an objective basis.

This book is the first of two volumes concerned with the analysis of the major groups of plant hormones, namely abscisic acid, cytokinins, ethylene, gibberellins, and indole-3-acetic acid. This first volume contains an introductory article on the principles of plant hormone analysis (Crozier) together with comprehensive chapters on practical aspects associated with the analysis of gibberellins (Hedden) and abscisic acid and related compounds (Neill and Horgan). Details are provided on the extraction of these compounds from plant tissues together with information on partitioning steps and purification techniques that can be used to achieve adequate levels of sample purity prior to analysis. Consideration is then given to the use of procedures, such as high-performance liquid chromatography, immunoassay, gas chromatography and combined gas chromatograpy–mass spectrometry, operating in the full scan and multiple ion monitoring modes, for qualitative and quantitative analyses.

Both this book and its sister volume contain a wealth of information that will be of unrivalled value as authoritative texts and comprehensive laboratory guides for day-to-day reference by those with interests in endogenous plant hormones whether it be from a physiological, chemical, biochemical or molecular biological view point. The books will also be of value to those with more general interests in analytical chemistry, as the techniques that are described and the philosophy underlying the design of analytical protocols are of relevance to the analysis of almost all naturally occurring organic compounds.

We are very grateful to the authors who researched their articles so thoroughly and responded so promptly to discussion and editorial comment. Their articles reflect their considerable first-hand experience and appreciation of the subtle and not-so-subtle ambiguities of plant hormone analysis. Finally, thanks are also due to Alison Sutcliffe and Norman Tait, Department of Botany, University of Glasgow who respectively redrew and photographed a number of the original diagrams supplied by the authors.

Alan Crozier
Department of Botany
The University
Glasgow
UK

Laurent Rivier
Institut Universitaire de Médecine Légale
Laboratoire de Toxicologie Analytique
Lausanne
Switzerland

Contents of Volume 1

Contents of Volume 2

6
Ethylene

1

Plant Hormone Analysis: Theoretical Considerations

Alan Crozier

I. INTRODUCTION

From a simple phytochemical view, plant hormones can be regarded as very minor secondary metabolites. Plants produce a vast array of secondary metabolites and, as a consequence, plant extracts, which are the starting point for hormone analysis, are exceedingly complex multicomponent mixtures. Hormones are typically present in only trace amounts and this makes accurate qualitative and quantitative analysis a difficult task. In the last decade, however, there have been rapid advances in the analytical sciences and many of the techniques that have come into routine use can be applied successfully to the purification, identification and measurement of nanogram quantities of hormones and related compounds in plant extracts. In fact, the large number of diverse methods that are currently available makes the choice of suitable procedures a daunting prospect if investigators are not fully acquainted with the practical and theoretical basis of the analytical problems with which they are confronted.

Reeve and Crozier (1980) have proposed that analytical methods can be classified according to whether they are based on selectivity or information. Systems of high selectivity tend to offer superior limits of detection and have a low information content, while those of low selectivity usually exhibit an inferior detection limit and can, although do not always, yield a greater amount of information. This principle can be seen to be operating in the case of a mass spectrometer. For instance, in order to obtain a 50–550 m/z mass spectrum, which provides an informative description of the sample under study, *ca.* 10 ng of material is a typical minimum requirement. When used in this mode the selectivity of the mass spectrometer is very low, as virtually all

The Principles and Practice of Plant Hormone Analysis
0-12-198375-7

organic compounds will evoke a response. However, if only one *m*/*z* value is monitored, the response is restricted to compounds yielding fragment ions at that *m*/*z* value and, as a consequence, selectivity and the limit of detection are enhanced markedly while information is reduced 500-fold.

In theory the distinction between qualitative and quantitative analysis is a semantic convenience rather than a logical reality. Nevertheless, there are fundamental differences in the practical approach that is taken as, in general, qualitative analysis employs techniques with a high information content while quantitative estimates are obtained with selective procedures.

II. QUALITATIVE ANALYSIS

There are two forms of qualitative analysis. In the first, identification is based on a direct comparison of the properties of the unknown, typically its mass spectrum, with those of a reference compound. The second method applies to the identification of new compounds and involves the deduction of the chemical structure of the unknown from its spectroscopic and chemical properties. The proposed structure is then synthesized from a known compound and the identity of the unknown confirmed by comparing its properties with those of the synthesized standard. In both instances characterization is based on the information obtained from the analysis and, because the techniques utilized (mass, IR, UV and NMR spectrometry, X-ray crystallography, etc.) are non-selective, sample purity is essential. The amount of information that is generated pertaining to the compound under study, and hence the accuracy of the analysis, is related closely to the amount of sample that is available. Depending upon the concentration of hormone in the tissue, anything from gram to kilogram quantities of plant material are required for extraction, and purification is achieved most effectively by using a combination of procedures in which the individual methods have distinctly different separatory mechanisms.

Reeve and Crozier (1980) have suggested that both the information content of individual procedures and the amount of information accumulated during an analysis, relating to the compound under investigation, can be defined in terms of the binary digit or bit. There has subsequently been much debate about these proposals, and doubt has been expressed about their relevance because of the difficulties in distinguishing between correlated and non-correlated information. There has also been debate on the relative merits of information-based identifications relying on chromatographic rather than mass-spectrometric evidence. Most investigators have an intuitive preference for mass spectrometric-based analyses and consider the data superior to those obtained by chromatographic techniques. This is because mass spectra are more understandable in terms of structural concepts, with fragmentation patterns being readily reduced to fit with our beguiling (though abstract) picture of atoms, bonds and molecules. Readers not familiar with this controversy should evaluate carefully the finer details of the logic of Reeve and Crozier (1980, 1983), Scott (1982, 1983) and MacMillan (1984, 1985).

III. QUANTITATIVE ANALYSIS

Quantitative analysis tends to make use of selective rather than informative techniques. There is thus a marked lowering of detection limits, which facilitates the analysis of extracts from relatively small amounts of tissue. Because of the selectivity of the procedures, many impurities go undetected and do not interfere with the analysis. The emphasis on purification is, thus, less severe and there is a resultant increase in the speed at which samples can be analysed.

A variety of selective techniques is available for quantitative analysis of plant hormones. The most ubiquitous is combined gas chromatography–mass spectrometry (GC–MS) operating in the multiple ion monitoring (MIM) mode. It is common practice to monitor four ions—two characteristic fragments in the spectrum of the endogenous compound of interest, frequently the molecular ion and base peak, and the equivalent ions in the spectrum of the ^{2}H-, ^{13}C-, ^{15}N- or ^{18}O-labelled analogue that is used as an internal standard. Quantitative estimates are then based on the relative responses at the appropriate GC retention time. GC–MIM is extremely flexible as the *m*/*z* values that are monitored can be changed to provide selective detection of numerous compounds.

Provided appropriate antibodies have been raised, immunoassays represent an alternative selective detector system for a number of plant hormones (see Weiler, 1984), although the basis of the selectivity is quite distinct from that obtained by GC–MIM. Selective analysis of indole-3-acetic acid (IAA) can also be achieved by using either the 2-methylindolo-α-pyrone assay (Stoessl and Venis, 1970), GC with a nitrogen–phosphorus detector (NPD) (Swartz and Powell, 1979; Martin *et al.*, 1980) or high-performance liquid chromatography (HPLC) with a fluorescence or electrochemical detector (Sweetser and Swartzfager, 1978). Selective GC analysis of cytokinins, abscisic acid and ethylene can be obtained with a NPD, electron capture and photo-ionization detectors respectively (Zelleke *et al.*, 1980; Seeley and Powell, 1970; Bassi and Spencer, 1985).

The presence of impurities which interfere with the analysis is usually self-evident when techniques with a high information content are utilized.

However, they are not recognized so readily in analyses based on selective procedures. As a consequence, it is important to take steps to verify the accuracy of quantitative estimates. In practice, this means taking precautions to ensure that there is a low probability of impurities contributing significantly to the quantified response.

Verification of accuracy is achieved readily, on a routine basis, with GC–MIM as abnormal *m/z* response ratios provide clear evidence of the presence of impurities. Unobserved interference will only occur when an impurity not only co-chromatographs with the hormone of interest but also induces a detector response at the *m/z* values monitored that is indistinguishable from that of either the compound of interest or the internal standard. The selectivity of the detector and peak capacity of capillary GC are such that the probability of such an occurrence is very low.

When selective procedures other than GC–MIM are employed for quantitative analysis, verification of accuracy becomes a more complicated process. One approach is to apply a "successive approximation", as outlined by Reeve and Crozier (1980). This involves analysing the sample and obtaining an estimate (E_1) of the amount of hormone present. The sample is then purified at least twofold and the hormone content (E_2) re-estimated. If E_1 is accurate, E_2, taking into account sample recovery and the precision of the method, should not be significantly different. If a difference is found, E_1 must be rejected as inaccurate and E_2 retested by further purification and analysis. This process is continued for as long as is necessary to obtain an estimate that does not change upon purification. At this point it becomes possible to conclude that on the basis of the available evidence there are no grounds for believing the estimate to be incorrect.

It is frequently not a practical proposition to carry out a formal successive approximation with every sample. In these circumstances one compromise that can be adopted is to carry out successive approximations on a number of typical extracts from the tissue of interest in order to determine the purification steps that are required before accurate estimates can be obtained. The procedures are then used to purify samples on a routine basis prior to quantitative analysis. It is, however, a useful precaution to "over-purify" extracts in order to reduce the possibility of inaccurate estimates being obtained when "rogue" samples are investigated.

The amount of sample purification that is a prerequisite for accurate analysis can also be assessed by using two distinctly different methods of selective detection. Accurate analysis is indicated when the two procedures provide estimates that are not significantly different. This approach was taken by Pengelly *et al.* (1981) when radioimmunoassay estimates of the IAA content of extracts of etiolated *Zea mays* shoots were validated by comparison with data obtained by GC–MIM.

It is very important to appreciate that each tissue contains not only varying amounts of hormones but also, in far greater quantities, its own characteristic assortment of impurities. As a consequence, when investigating an endogenous hormone in a particular plant material, it is essential to evaluate purification and analytical procedures, as outlined above, before beginning routine quantitative analysis. The application of standardized recipes to extracts from diverse sources, without prior experimentation, is an approach that sooner, rather than later, will result in the acquisition of inaccurate data.

Once routine analyses are in progress further checks on accuracy can often be undertaken. For instance, when HPLC is used with either an absorbance monitor or a fluorimeter, details of absorbance or fluorescence activation and emission spectra provide a simple check on peak homogeneity. Estimates obtained with immunoassays or the 2-methylindolo-α-pyrone assay can be checked by assaying a range of sample dilutions. Parallel curves indicate an absence of interference. In addition, equal-sized aliquots can be assayed together with increasing amounts of standard. The hormone measured is then plotted against standard added. The curve should be a straight line with unit slope and a y-intercept equal to the amount of endogenous hormone in the sample. The presence of interfering compounds which affect accuracy adversely is usually indicated by a change in the slope of the curve (Pengelly and Meins, 1977; Crozier *et al.*, 1986; Pengelly, 1986).

IV. ANALYSIS OF ISOTOPICALLY-LABELLED METABOLITES

In the analysis of endogenous hormones, the compound of interest must be distinguished from a vast number of organic compounds. However, when HPLC or GC are used with radioactivity counting (RC) to investigate ^{3}H- or ^{14}C-labelled products of metabolism experiments, the analytical situation is simplified greatly. This is because once minimal sample purification has been achieved unlabelled endogenous components do not influence the analysis, and metabolites have to be distinguished from a very restricted population of radiolabelled compounds. This does not apply when GC–MIM is used to analyse isotopically-labelled metabolites, because ions originating from unlabelled impurities can interfere with the analysis. Thus, although the detector provides high selectivity, the analytical situation is not simplified, as it is with HPLC–RC or GC–RC, as metabolites have to be identified against the background of an extremely complex mixture of organic constituents.

ACKNOWLEDGEMENTS

The author would like to thank Dr Roger Horgan, Ms Joan Malcolm and Dr Ana Maria Monteiro for their invaluable assistance in the preparation of this article.

REFERENCES

Bassi, P. K. and Spencer, M. S. (1985). Comparative evaluation of photoionization and flame ionization detectors for ethylene analysis. *Plant Cell Environm.* **8**, 161–165.

Crozier, A., Sandberg, G., Monteiro, A. M. and Sundberg, G. (1986). The use of immunological techniques in plant hormone analysis. *In* "Plant Growth Substances 1985" (M. Bopp, ed.) pp. 13–21. Springer-Verlag, Berlin.

MacMillan, J. (1984). Analysis of plant hormones and metabolism of gibberellins. *In* "The Biosynthesis and Metabolism of Plant Hormones", Society for Experimental Biology Seminar Series 23 (A. Crozier and J. R. Hillman, eds), pp. 1–16. Cambridge University Press, Cambridge.

MacMillan, J. (1985). Gibberellin metabolism: objectives and methodology. *Biol. Plant.* **27**, 164–171.

Martin, C. G., Nishijima, C. and Labavitch, J. M. (1980). Analysis of indoleacetic acid by the nitrogen-phosphorous gas chromatograph. *J. Amer. Soc. Hort. Sci.* **105**, 46–50.

Pengelly, W. (1986). Validation of immunoassays. *In* "Plant Growth Substances 1985" (M. Bopp, ed.) pp. 35–43. Springer-Verlag, Berlin.

Pengelly, W. and Meins, F. (1977). A specific radioimmunoassay for nanogram quantities of the auxin, indole-3-acetic acid. *Planta* **136**, 173–180.

Pengelly, W., Bandurski, R. S. and Schulze, A. (1981). Validation of a radioimmunoassay for indole-3-acetic acid using gas chromatography-selected ion monitoring-mass spectrometry. *Plant Physiol.* **68**, 96–98.

Reeve, D. R. and Crozier, A. (1980). Quantitative analysis of plant hormones. *In* "Hormonal Regulation of Development. 1. Molecular Aspects of Plant Hormones", Encyclopedia of Plant Physiology, New Series, Vol. 9 (J. MacMillan, ed.) pp. 203–280. Springer-Verlag, Berlin.

Reeve, D. R. and Crozier, A. (1983). A reply to "Information theory and plant growth substance analysis" by I. M. Scott. *Plant Cell Environm.* **6**, 365–367.

Scott, I. M. (1982). Information theory and plant growth substance analysis. *Plant Cell Environm.* **5**, 339–342.

Scott, I. M. (1983). An answer to Reeve and Crozier's reply. *Plant Cell Environm.* **6**, 367–368.

Seeley, S. D. and Powell, L. E. (1970). Electron capture-gas chromatography for sensitive assay of abscisic acid. *Anal. Biochem.* **35**, 530–533.

Stoessl, A. and Venis, M. A. (1970). Determination of submicrogram levels of indole-3-acetic acid: A new highly specific method. *Anal. Biochem.* **34**, 344–351.

Swartz, H. J. and Powell, L. E. (1979). Determination of indoleacetic acid from plant tissues by an alkali flame ionization detector. *Physiol. Plant.* **47**, 25–28.

Sweetser, P. B. and Swartzfager, D. G. (1978). Indole-3-acetic acid levels in plant tissues as determined by a new high performance liquid chromatographic method. *Plant Physiol.* **61**, 254–258.

Weiler, E. W. (1984). Immunoassay of plant growth regulators. *Annu. Rev. Plant. Physiol.* **35**, 85–95.

Zelleke, A., Martin, G. C. and Labavitch, J. M. (1980). Detection of cytokinins using a gas chromatograph equipped with a sensitive nitrogen–phosphorous detector. *J. Am. Soc. Hort. Sci.* **105**, 50–53.

2

Gibberellins

Peter Hedden

I. INTRODUCTION

There are a number of technical difficulties associated with the analysis of gibberellins (GAs) in higher plants. In common with most other classes of plant growth substances the concentrations of GAs in plant tissues can be exceedingly low, especially in vegetative material. Thus GA analysis requires very sensitive methods of detection. Furthermore, the analysis is complicated by the large number of different GAs that may be encountered. To date, 72 GAs have been characterized from higher plants and fungi, and individual species can contain several different GAs. Thus analytical procedures used to investigate endogenous GAs must be able to distinguish a GA from the millions of other possible plant components, as well as from other GAs from which it may differ only very slightly. Furthermore, the problem is compounded because the GAs display virtually no characteristics, such as fluorescence or strong UV absorption, that might easily distinguish them from other organic acids. There are, therefore, few methods available that are both sufficiently sensitive and sufficiently selective to give reliable estimates of GA concentrations in small samples of tissue.

Although bioassays are useful for establishing the presence of GAs in extracts and for tracing these substances throughout a purification procedure, it is now generally appreciated that they lack precision and, because of the influence of contaminants, they give unreliable estimates of GA concentrations. GA bioassays are therefore of limited value as analytical techniques and thus they are not covered in this chapter.

The Principles and Practice of Plant Hormone Analysis
0-12-198375-7

The multiplicity of GAs within a single plant species has provided plant physiologists with a dilemma. Even if we concern ourselves only with those molecules that exhibit biological activity, interest is still focused on several GAs within a single plant organ. In attempting to correlate plant developmental changes with differences in hormone concentration we are often obliged to consider simultaneous changes in a number of GAs, thus complicating both the analysis and the interpretation of the results. Recent ideas stemming largely from work with GA-deficient mutants have sought to restrict the number of physiologically important GAs by suggesting that most GAs are either metabolic precursors or deactivation products of the true hormone (Phinney, 1984; Potts and Reid, 1983). Many GAs also appear to be formed via routes that branch from the pathway to the active compound. These GAs may have little physiological function. Thus the analytical problem might be simplified, provided we know which GAs are important. This may not be determined until considerably more is known about the mechanism of GA action. In any case there is unlikely to be a universal "active" GA, as at least some differences between species are to be expected.

Another problem for plant physiologists is to ensure homogeneity within a sample population. The limitations on the sensitivity of available analytical methods have necessitated bulking many individual plants or tissues for extraction and analysis. Any scatter in developmental stage or organ composition among these individuals will tend to make differences in hormone concentration between samples less distinct. In order to overcome this problem we should ideally extract single plant organs of a carefully defined developmental stage. Modern analytical methods such as combined gas chromatography–mass spectrometry (GC–MS) and immunoassay are now becoming sufficiently sensitive to allow such analyses to be made, and should lead to significant advances in our understanding of hormone physiology. Ultimately we may need to examine the growth substance content of individual cells or organelles.

It is not the purpose of this chapter to provide a review of the literature on analytical methods for GAs. Rather it is intended to be a guide to the design of procedures for GA analyses and, to that end, will discuss the options available. The reader's attention is drawn to some recent articles that give extensive coverage to this subject (Graebe and Ropers, 1978; Yokota *et al.*, 1980; Reeve and Crozier, 1980; Horgan, 1981; Crozier and Durley, 1983). It should be emphasized that no universal method exists for the analysis of GAs in plant tissues, but that procedures should fit the requirements of the problems presented by each case. Factors such as the concentration of the GAs and the nature of the contaminants play an important part in determining which procedures should be used.

II. CHEMISTRY

A. Structure and biosynthesis

The GAs form a large group of tetracyclic diterpenoid carboxylic acids based on the *ent*-gibberellane skeleton (**1**). Over 80 structurally different GAs have been isolated from higher plants and fungi, and of these 72 have been sufficiently well characterized to warrant an A-number (MacMillan and Takahashi, 1968). The structures of these GAs are given in Fig. 1.

The GAs can be divided into two main groups, the C_{20}-GAs as typified by the simplest member, GA_{12}, and the C_{19}-GAs such as GA_9. The C_{20}-GAs are known to be the metabolic precursors of the C_{19}-GAs, which include the physiologically active hormones. The C_{20}-GAs can be subdivided into four groups differing in the degree of oxidation at C-20. Again these structural

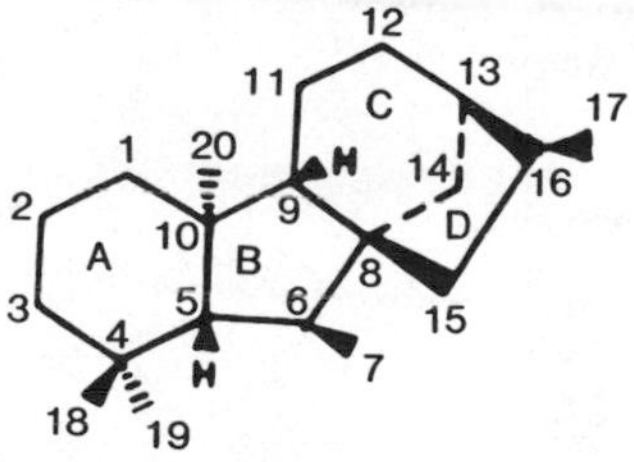

1 *ent*-gibberellane

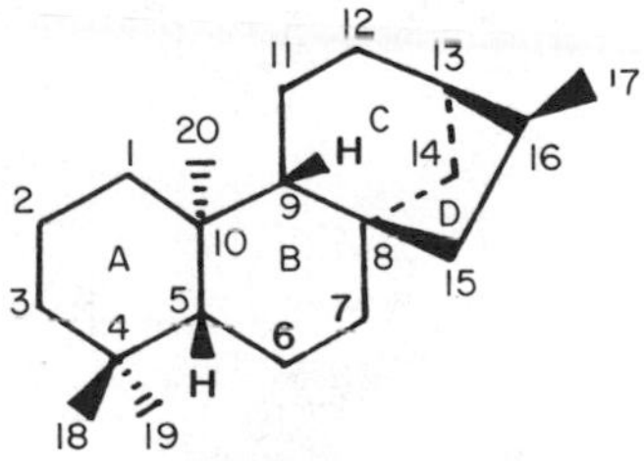

2 *ent*-kaurane

types are metabolically related as shown by the biosynthetic sequence in Fig. 2, which is based on data obtained with cell-free preparations from *Pisum sativum* by Kamiya and Graebe (1983). δ-lactones such as GA_{44} are now thought to be artefacts produced during extraction at low pH. The true intermediate probably has a free hydroxyl group at C-20 as depicted in Fig. 2 (Hedden and Graebe, 1982; Kamiya and Graebe, 1983). Formation of the C_{19}-GAs occurs by loss of C-20 at the aldehyde oxidation level by an as yet unknown mechanism. All C_{19}-GAs contain a γ-lactone, usually between C-19 and C-10, and, with two exceptions, are monocarboxylic acids. Within each of these groups there is considerable variation in the degree and position of oxidation, which is usually hydroxylation. The major positions of hydroxylation are indicated in Fig. 3. Hydroxylation at the 3β- and 13-positions is particularly common, as is 2β-hydroxylation, which results in loss of biological activity and is thought to be a deactivating mechanism.

The biosynthetic pathway from mevalonic acid (MVA) to the GAs is outlined in Fig. 4. The immediate precursors of GAs are *ent*-kaurenoid compounds, which are structurally based on *ent*-kaurane (**2**). Figure 4 also

GA_1 GA_2 GA_3

GA_4 GA_5 GA_6

GA_7 GA_8 GA_9

GA_{10} GA_{11} GA_{12}

GA_{13} GA_{14} GA_{15}

GA_{16} GA_{17} GA_{18}

GA_{19} GA_{20} GA_{21}

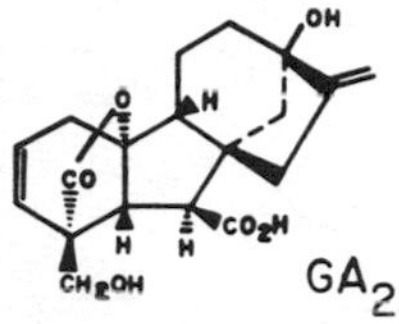

GA_{22} 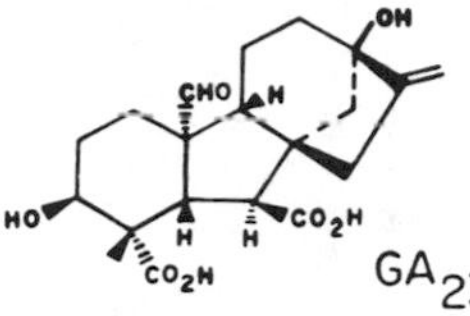 GA_{23} 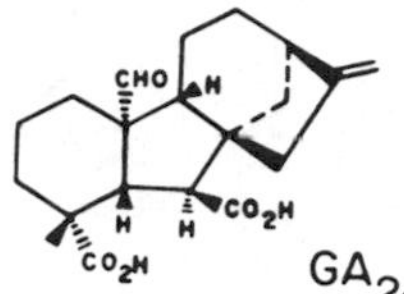 GA_{24}

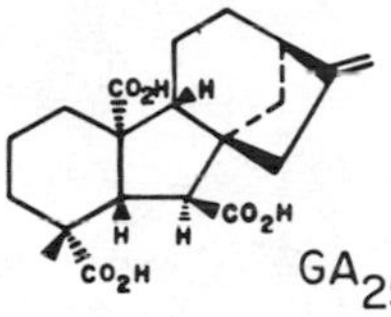

GA_{25} 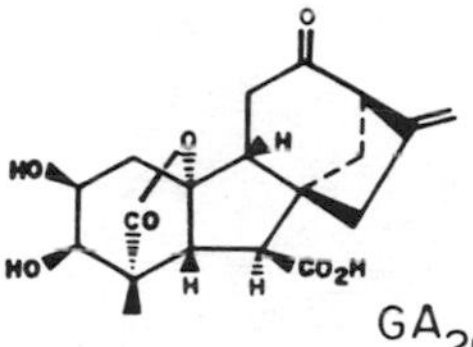 GA_{26} 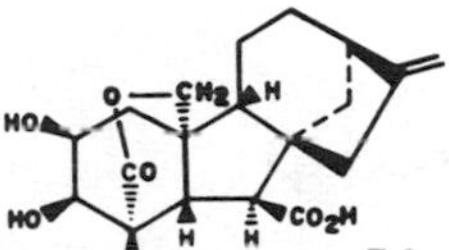 GA_{27}

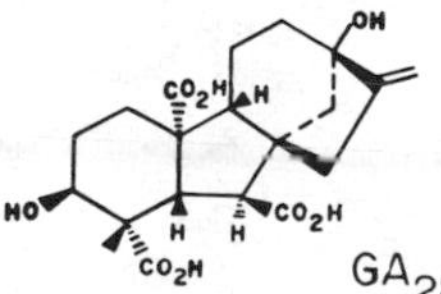

GA_{28} 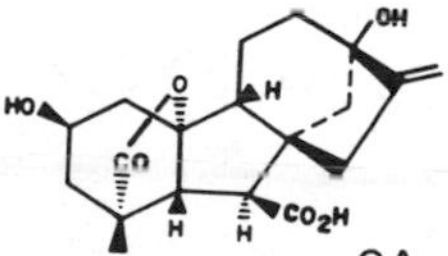 GA_{29} 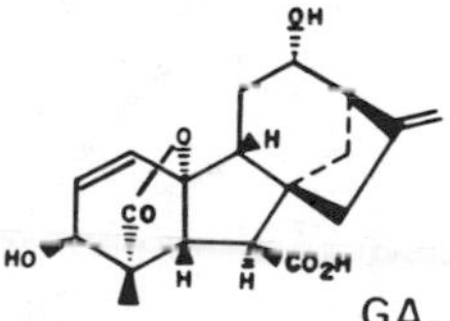 GA_{30}

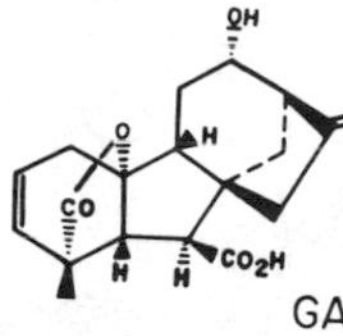

GA_{31} 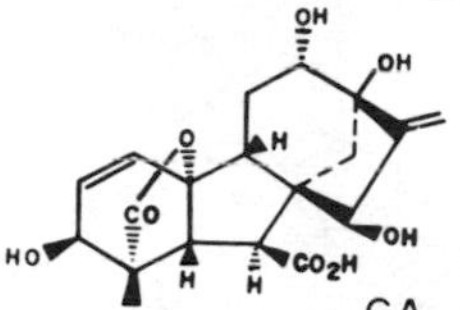 GA_{32} 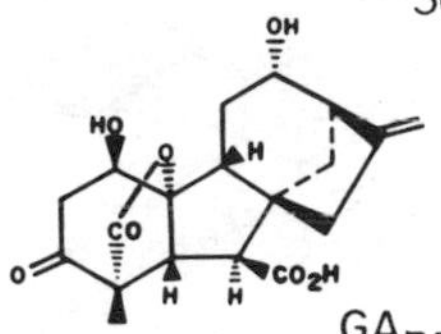 GA_{33}

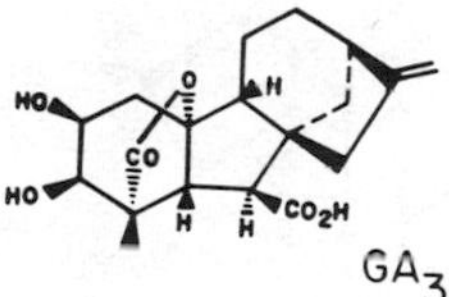

GA_{34} 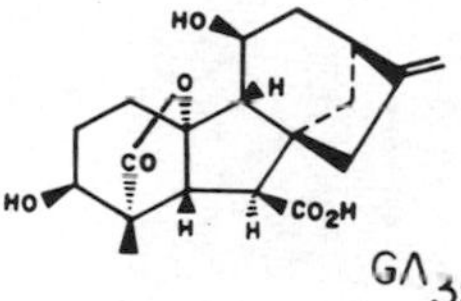 GA_{35} 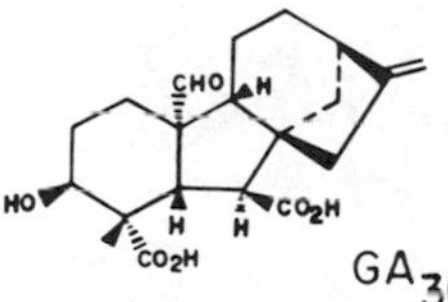 GA_{36}

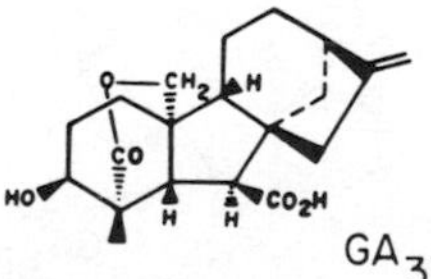

GA_{37} 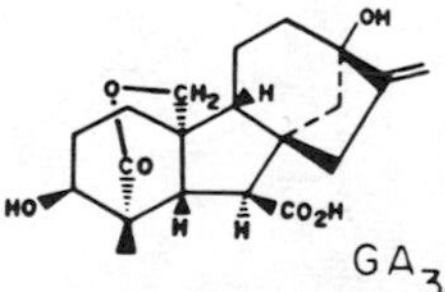 GA_{38} 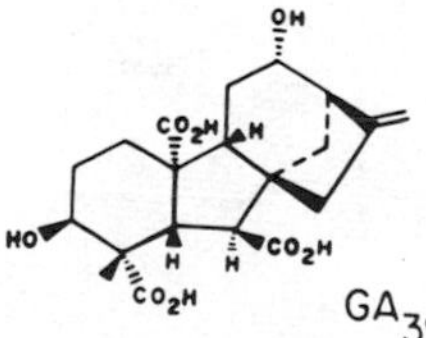 GA_{39}

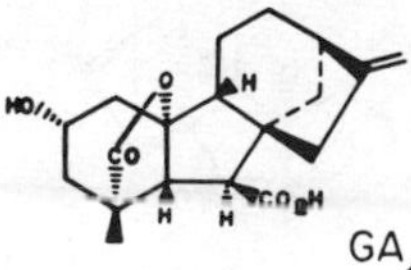

GA_{40} 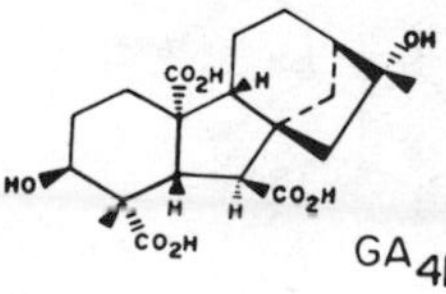 GA_{41} 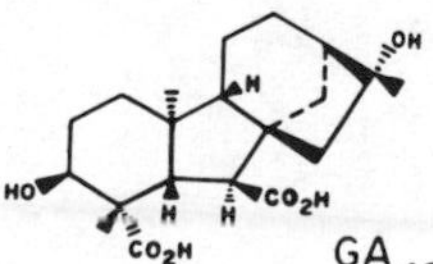 GA_{42}

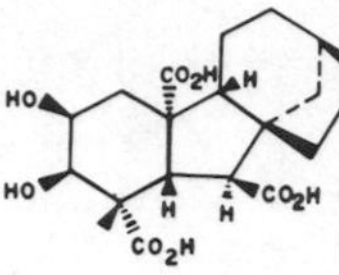

GA_{43} 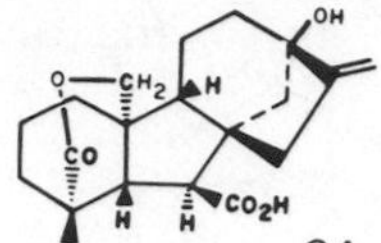 GA_{44} 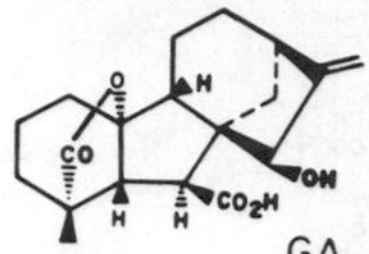 GA_{45}

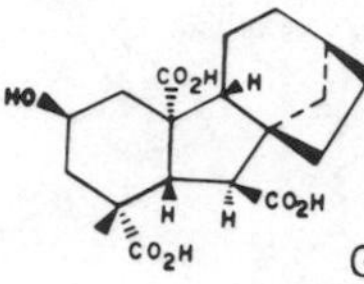

GA_{46} 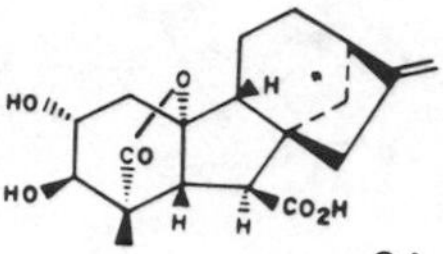 GA_{47} 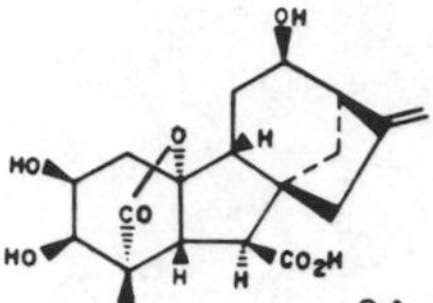 GA_{48}

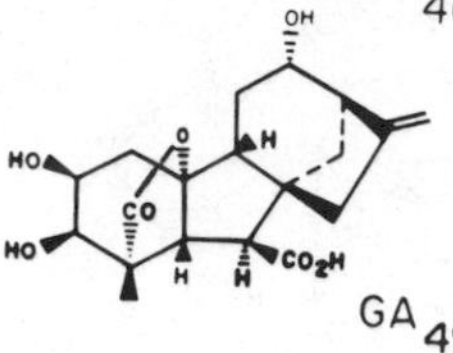

GA_{49} 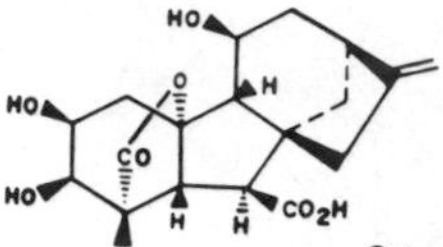 GA_{50} 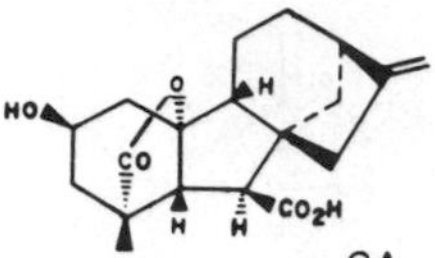 GA_{51}

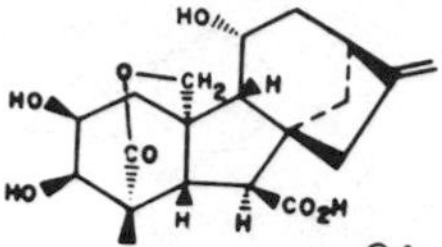

GA_{52} 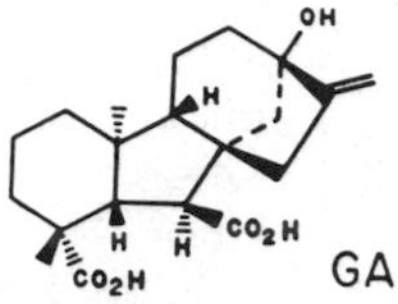 GA_{53} 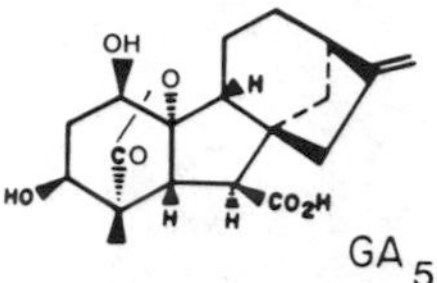 GA_{54}

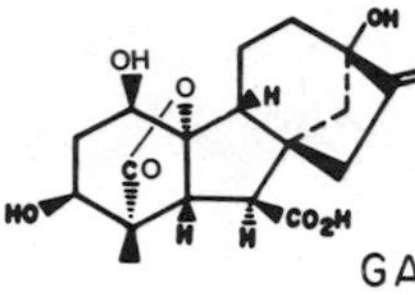

GA_{55} 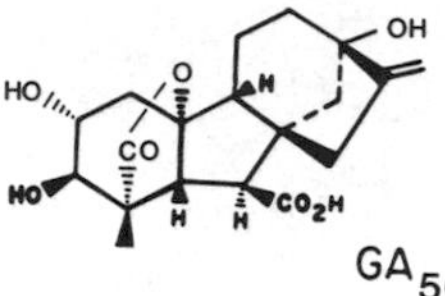 GA_{56} 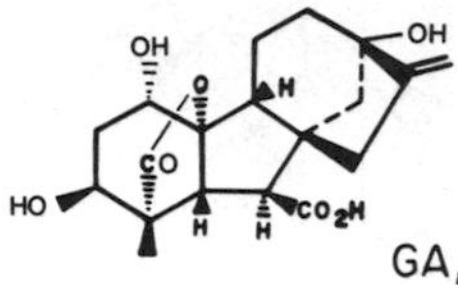 GA_{57}

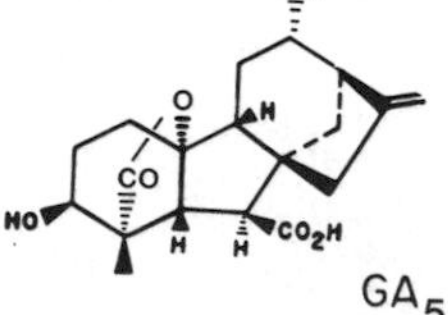

GA_{58} 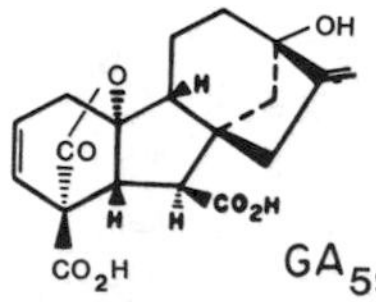 GA_{59} 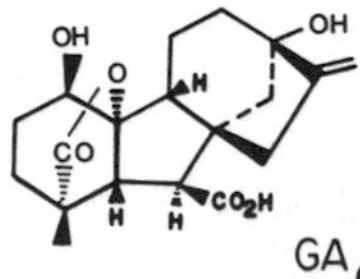 GA_{60}

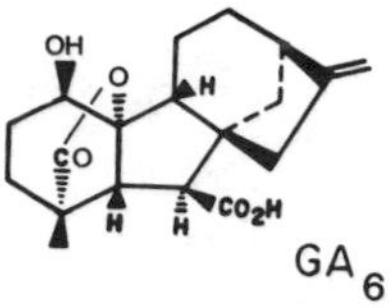

GA_{61} 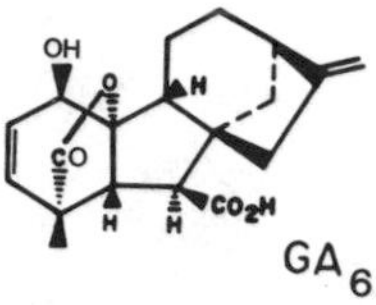 GA_{62} 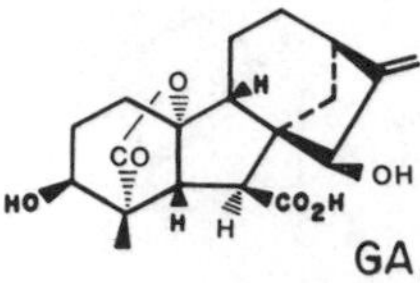 GA_{63}

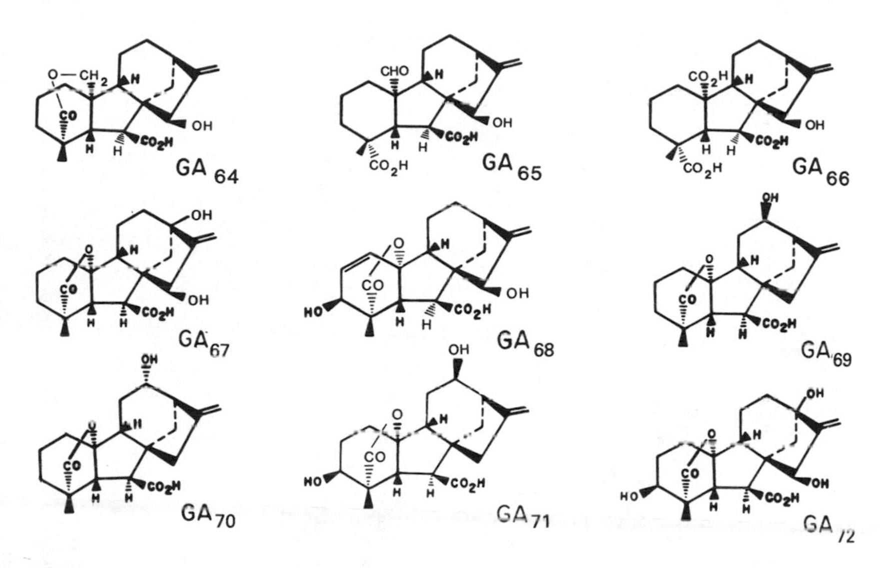

Fig. 1. Structures of the characterized gibberellins. The structures of GAs 1-52 are reproduced, with permission, from the Annual Review of Plant Physiology, Vol. 29. © 1978 by Annual Reviews Inc.

shows pathways to some metabolically-related *ent*-kaurenoids that often accompany GAs in plant or fungal extracts (Hedden, 1983). The *ent*-kaurenoids are named semisystematically. Note that the presence of the prefix "*ent*" reverses stereochemical assignments, i.e. $ent\text{-}\alpha \equiv \beta$ (Rowe, 1968). Thus the full nomenclature of the immediate precursor of GA_{12}-aldehyde is *ent*-7α-hydroxykaur-16-en-19-oic acid (Fig. 4). The same applies to the GAs when they are referred to by the systematic *ent* gibberellane (**1**) nomenclature (Rowe, 1968). The trivial names $GA_{1...n}$ are used much more frequently, however, and in these circumstances it is not necessary to reverse the stereochemical assignments.

B. Stability

The physiologically important GAs are usually highly oxidized molecules with many functional groups. As a consequence they tend to be quite labile, especially in aqueous solution at extremes of pH and at elevated temperatures. Some of the more important rearrangements and degradations undergone by 3β- and 13-hydroxylated GAs are outlined in Fig. 5.

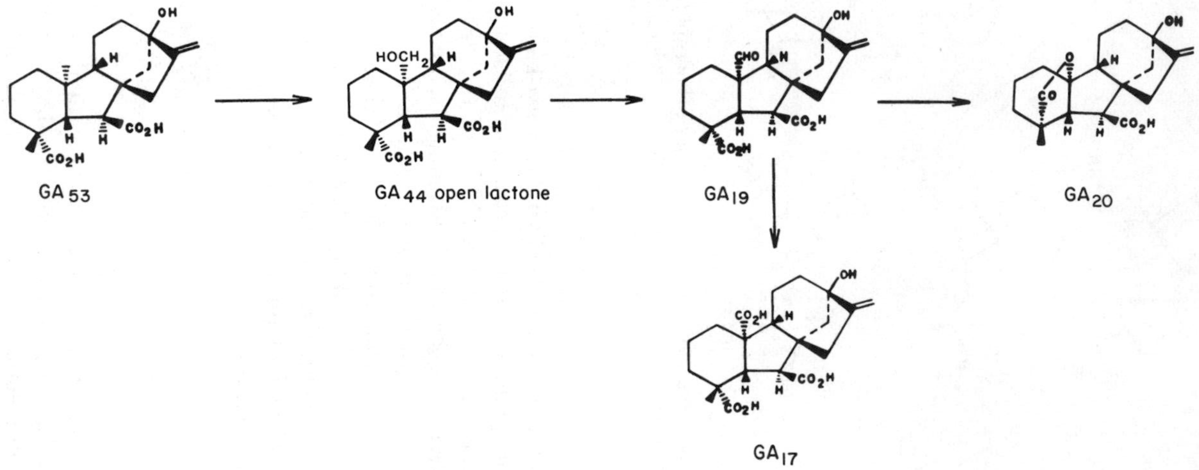

Fig. 2. Metabolic relationship between C_{20}- and C_{19}-GAs. This example is taken from the 13-hydroxylated GA pathway, which is known to occur in many higher plants.

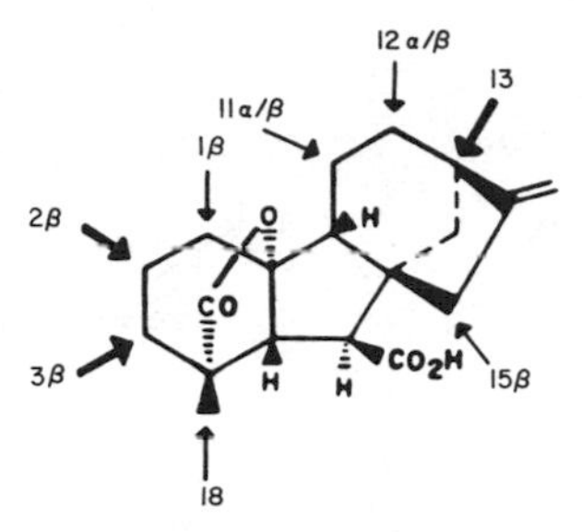

Fig. 3. The main positions of GA hydroxylation in higher plants. The thick arrows indicate the most common positions.

Under alkaline conditions GAs containing only a 3β-hydroxyl group in ring A undergo a reversible retro-aldol rearrangement, which results in epimerization (MacMillan and Pryce, 1973). GAs, such as GA_3 or GA_7, that contain a 1,2 double bond, isomerize under basic conditions to the 19-2 lactones with a shift of the double bond to the 1,10 position (Kirkwood *et al.*, 1980). This rearrangement occurs very easily, even in the heated injection port of a gas chromatograph. 13-Hydroxylated GAs are susceptible to a Wagner–Merwein rearrangement of the C/D ring system in mineral acid (Grove *et al.*, 1960). In the absence of a 13-hydroxyl group, low pH can result in hydration of the 16,17 double bond or its shift to the 15,16 endocyclic position. The combination of functional groups in GA_3 renders it particularly vulnerable to degradation. After heating in aqueous solution at 120°C for 20 minutes in an autoclave, only 1–2% of the original substance remains (Pryce, 1973).

It is clear from the foregoing discussion that extraction and purification procedures should avoid extremes of pH. It is advisable to remain within the range pH 2.5–8.5. Solutions containing GAs, especially aqueous solutions, are best handled at low temperatures, preferably below 40°C. Extracts or purified fractions therefrom are best stored in a freezer at −20°C. Certain GA biosynthetic precursors such as GA_{12}-aldehyde or *ent*-kaurene are prone to aerial oxidation when present in non-crystalline form. Although such samples may be kept for short periods at −20°C, they can be stored more safely under liquid nitrogen.

III. EXTRACTION AND PURIFICATION

As a consequence of the low concentration of GAs in plant tissues, elaborate purification procedures are usually necessary to yield preparations sufficiently pure for analysis. Many such procedures have been reported, their

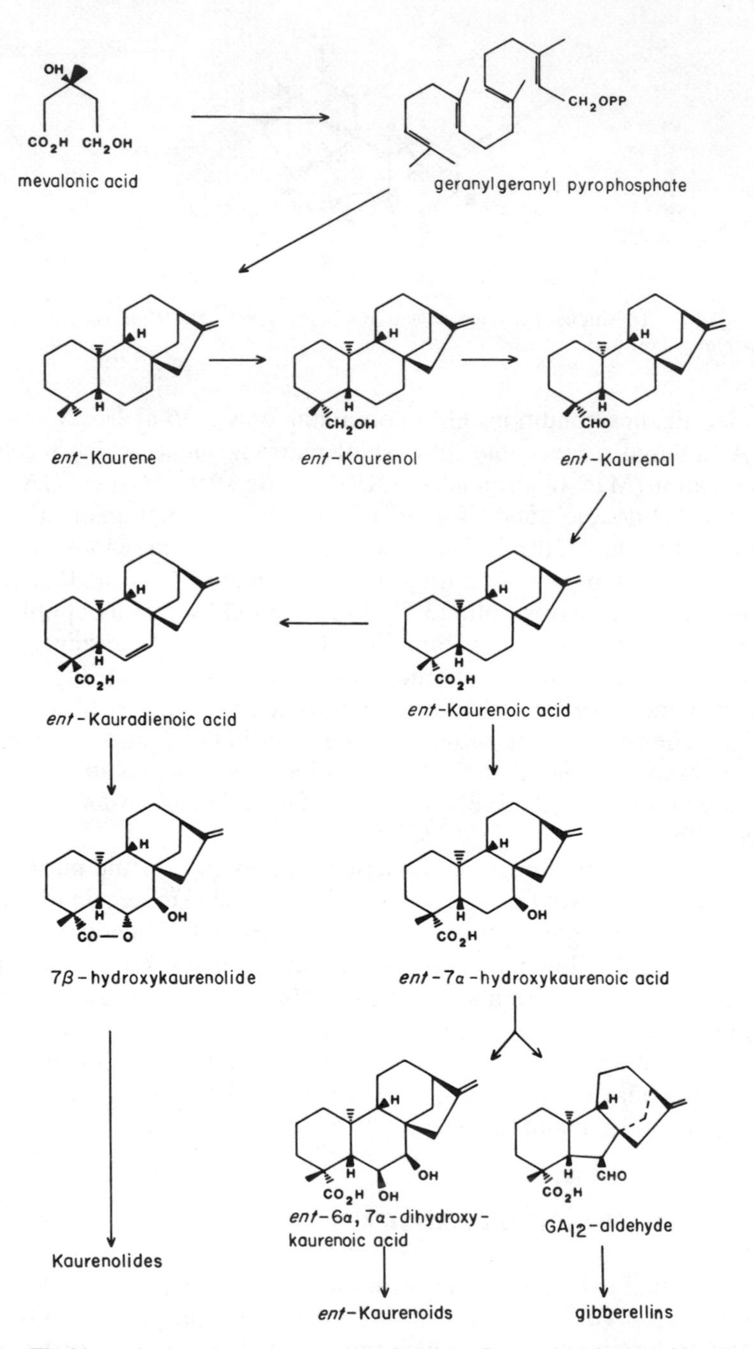

Fig. 4. The biosynthetic pathway from mevalonic acid to GAs and *ent*-kaurenoid compounds.

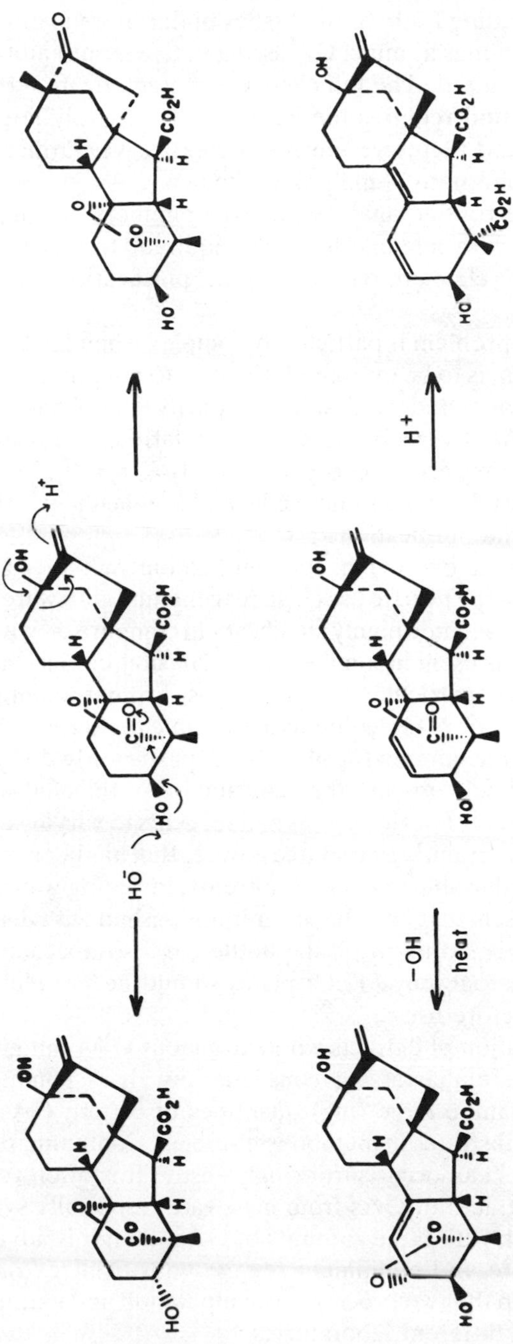

Fig. 5. Common rearrangements undergone by 3β- and 13-hydroxylated GAs.

complexity depending both on the tissues under investigation (seeds often contain up to 10^3 times as much GA as vegetative tissue) and on the degree of final purity required. The full chemical characterization of a previously unknown compound requires the isolation of relatively large amounts of pure substance and therefore much more extensive purification than, for example, the quantitative analysis of known GAs in small samples by GC–MS. For the routine analyses of large numbers of samples a simple, rapid procedure with minimal losses is required. In this regard immunoassays, which, it is claimed, require minimal purification, are potentially of great value.

The analytical problem is particularly complex when the full complement of GAs in a system is to be examined. In order to minimize the final number of fractions to be analysed it is desirable to purify the GAs as a group as far as possible. The GAs cover a broad range of polarities and the only property they share is that they are weak organic acids ($pK_a \approx 4.0$). Therefore simple partition of the crude extract into acidic and non-acidic fractions is usually the first step in the purification.

Before discussing extraction and purification methods in detail, it is important to stress the need to use clean reagents and glassware. Low levels of contaminants can become highly significant in comparison with the traces of endogenous GAs present in some extracts. The ubiquitous plasticizers, such as phthalates, are a particular problem in gas chromatography (GC) analyses since they elute from GC columns in the same region as GAs. In GC–MS analyses it is not uncommon for phthalates, particularly di-2-ethylhexylphthalate, to completely obscure the GAs and make the analysis impossible. Plasticizers are present in all solvents and reagents to varying extents (Ishida *et al.*, 1980). A particularly rich source is water that has been stored in plastic containers. It is advisable to distil all solvents, including water, and to store them in glass vessels that have been carefully washed with distilled solvent. Silica gel, which is stored in plastic bottles, is also a rich source, so that thin-layer chromatography (TLC) plates should be pre-eluted in a polar solvent system before use.

The contamination of extracts with extraneous GAs can give misleading results and have embarrassing consequences. It is common for many laboratories to handle large (mg) quantities of certain GAs in preparing standards or as substrates in metabolism studies. Contamination of extracts from such sources can occur surprisingly easily. It is often very difficult to remove the last traces of GAs from glassware, especially syringes. If it is necessary to work with large amounts of GAs, then it is advisable to keep separate glassware and equipment for use with plant extracts. The best solution is to keep the two processes (manipulation and extraction of GAs) quite separate in different laboratories.

A. Extraction

Extraction is usually accomplished by maceration in cold methanol or aqueous methanol (80% methanol). Acetone is occasionally used instead of methanol, but can lead to the formation of acetonides with geminal diols, especially at low pH (Gaskin and MacMillan, 1978). A 4–5-fold excess of extraction medium over tissue is generally sufficient, although when small amounts of tissue are extracted higher solvent/tissue ratios are often more practicable. The macerated tissue is agitated in the extracting solvent for several hours at low temperature (4°C), after which insoluble matter is removed, usually by filtration. The filtered residue is re-extracted or, at the very least, thoroughly washed with cold methanol. The methanol is removed from the combined extracts and washings by evaporation under reduced pressure and at a temperature that should not exceed 40°C.

Precipitated material, which can be increased by freezing and thawing the aqueous residue, can be removed by centrifugation. Removal of this material subsequently allows easier partitioning and produces a cleaner extract, although there is a danger that the more hydrophobic GAs and GA precursors may co-precipitate to some extent. At this stage the aqueous residue is often diluted with phosphate buffer (to give a final concentration of 50–100 mM) in order to increase the ionic strength of the solution. However, as will be discussed later, phosphoric acid, if allowed to concentrate, can lead to extensive degradation of GAs and some workers prefer to rely on the intrinsic, though slight, buffering capacity of the extract.

The extraction of GAs from fungal cultures is usually carried out by partition of the growth medium at pH 2.5 against ethyl acetate after the mycelia are removed by filtration (see for example Bearder *et al.*, 1975b). Although GAs are excreted into the growth medium, the more lipophilic precursors remain associated with the fungal mycelia from which they can be extracted with methanol.

GAs have been extracted from homogenized liquid endosperm of *Cucurbita maxima* by simply adjusting the homogenate to low pH and partitioning against ethyl acetate (Beale *et al.*, 1984). The use of phosphate buffer as the homogenization medium has been reported to give a cleaner extract than aqueous methanol (Jones, 1968). This is undoubtedly so, but this method could not be expected to extract lipophilic material very efficiently. In fact there is a report of GAs being extracted much more effectively with the detergent Triton X-100 than with methanol (Browning and Saunders, 1977). This indicates a close association of the GAs with membranes. Unfortunately there has been little corroborative evidence to support this finding.

The question of how efficiently GAs are extracted from plant tissue is difficult to answer. There is little firm information on the subcellular

B.

localization of GAs. It is likely that their extractability will be affected by factors such as association with lipids, phenolics or proteins, and will vary with the polarity of the molecule. Corrections cannot be made for this uncertainty in the degree of extraction, which is probably the largest source of error in estimates of GA concentrations.

B. Liquid–liquid partition procedures

Figure 6 outlines the crude separation of the extract into ethyl acetate-soluble acidic and neutral fractions and a butanol-soluble fraction. The vast majority of GAs, except for the most polar, will be recovered in the acidic fraction. Partition coefficients of GAs between phosphate buffer and ethyl

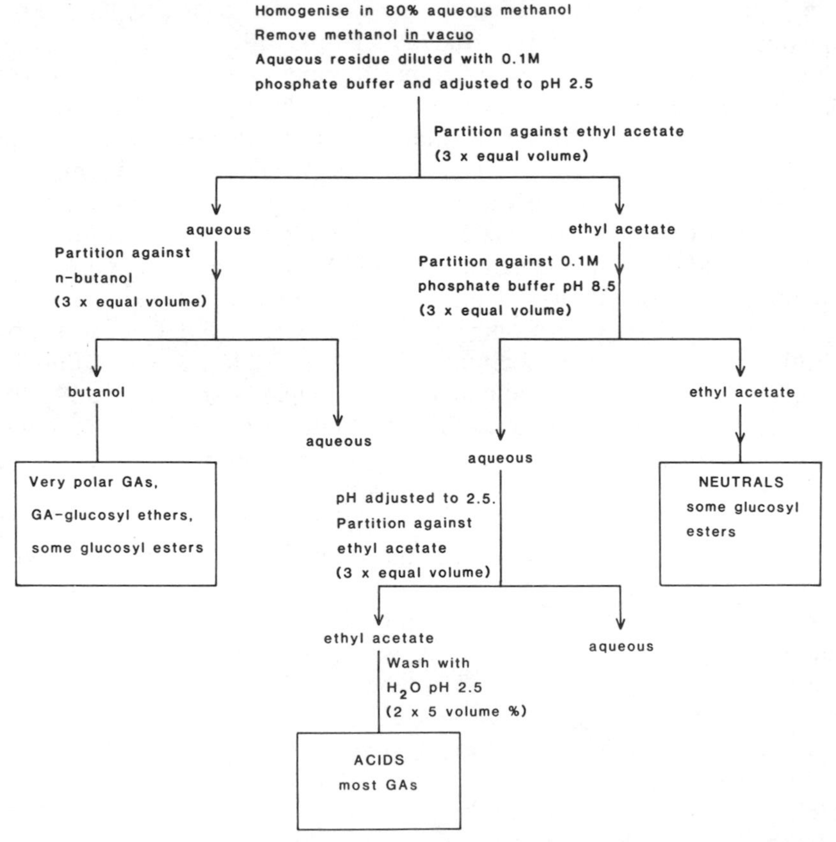

Fig. 6. Initial purification scheme for the separation of extracts into acidic, neutral/basic and 1-butanol-soluble (polar) fractions by solvent–solvent partition.

acetate have been published by Durley and Pharis (1972) and allow us to estimate recoveries. Table I, reproduced from Durley and Pharis's publication, shows the effect of pH and buffer molarity on the partition coefficients between phosphate buffer and ethyl acetate for four GAs. It can be calculated that at a final phosphate concentration of 0.05 M the recovery of GA_9 in the acidic ethyl acetate fraction is better than 99% and that of GA_3 is 86%. Of course these figures should be regarded only as indicative since plant extracts contain components that alter the partition coefficients.

Table I. The effect of buffer pH and molarity on the partition coefficients ($K_d = C_{aq}/C_{org}$) of GA_3, GA_5, GA_9 and GA_{13} between ethyl acetate and Na,K-phosphate.

		pH K_d				
Gibberellin	Molarity	8.0	6.5	5.0	3.5	2.5
A_3	1.5	∞	∞	1.2	0.21	0.17
	0.5	∞	∞	2.8	0.63	0.41
	0.1	∞	∞	7.2	1.4	0.86
	0.05	∞	∞	19.2	2.7	1.31
A_5	1.5	∞	4.8	0.19	0	0
	0.5	∞	14.2	0.41	0.02	0
	0.1	∞	∞	0.80	0.05	0
	0.05	∞	∞	1.2	0.09	0
A_9	1.5	0.34	0.06	0	0	0
	0.5	0.98	0.13	0	0	0
	0.1	2.9	0.29	0.02	0	0
	0.05	3.8	0.37	0.03	0	0
A_{13}	1.5	∞	7.1	0.06	0	0
	0.5	∞	20.0	0.13	0.02	0
	0.1	∞	∞	0.20	0.04	0
	0.05	∞	∞	0.28	0.08	0

Data taken from Durley and Pharis (1972).
Note: K_d values <0.02 are taken as 0; values >20 are taken as ∞.

The tetrahydroxy C_{19}-GA, GA_{32}, is too polar to partition into ethyl acetate and must be extracted at low pH with 1-butanol (Bukovac *et al.*, 1979). GA-glucosyl ethers also partition into the butanol fraction as do the more polar GA-glucosyl esters. Less polar glycosyl esters partition partly into the neutral ethyl acetate fraction. The analysis of GA conjugates will be discussed more fully in Section V.

As an alternative, the first aqueous phase can be partitioned against hexane or diethyl ether at high pH (8.5–9.0) to remove lipid material, and the GAs then extracted into ethyl acetate at pH 2.5. This method, although simpler than the scheme outlined in Fig. 6, is unlikely to give as efficient a purification. Initial partition against ethyl acetate at pH 8.5, a procedure that has also been employed, results in partial extraction of less polar GAs into the organic phase.

The presence of surface-active components in the crude extract can result in the formation of emulsions in the initial partition step. Several methods can be tried for breaking emulsions, such as filtration through celite, the addition of salts or a few drops of ethanol, but centrifugation is often the most effective procedure if the volume is not large.

Traces of acid in the acidic ethyl acetate phase become concentrated on evaporation of the solvent. This is particularly a problem with phosphoric acid, which is produced when phosphate buffer is acidified, since this acid is less volatile than water. Rademacher (1978) has shown that 0.1 mg phosphoric acid in 50 ml water-saturated ethyl acetate leads to more than 90% destruction of GA_3 and GA_4 when the solution is concentrated to dryness at 30°C under reduced pressure (Fig. 7*A*). Thus the ethyl acetate phase should be washed with water to remove the last traces of phosphoric acid before concentrating to dryness. In order to prevent partitioning of the GAs into the water it is acidified to pH 2.5 by the addition of a few drops of hydrochloric acid. Being a gas, this is quickly lost during evaporation of the solvent. Four washes with water at pH 2.5 (5% volume of the organic phase) results in virtually complete recovery of GA_3 and GA_4 after concentration of the organic phase to dryness. This is illustrated in Fig. 7*B*, which is reproduced from Rademacher (1978). Water-saturated ethyl acetate contains 5% water, which is sometimes removed with anhydrous sodium sulphate. This step can, however, lead to losses by adsorption, and, if washing with acidified water is carried out, it is unnecessary. Water remaining after evaporation of the ethyl acetate can be removed azeotropically by the addition of toluene and further evaporation.

C. Group separation methods

Separation of the crude extract into the ethyl acetate-soluble acid and neutral fractions does not usually reduce the weight of extract to within the sample capacity of most analytical procedures. One or more of a number of methods are routinely employed that, as far as possible, purify the GAs as a group. The major impurities remaining after liquid–liquid partition are fatty acids, phenolics and sugars. The last, although neutral and highly water-soluble, are often present in plant tissues in such high concentrations that small amounts may spill over into the acid fraction. Fatty acids, especially those that are hydroxylated, are chemically so similar to GAs that they cannot be effectively separated from them as a group. Phenolics present a particular problem since they form complexes with GAs that are often quite

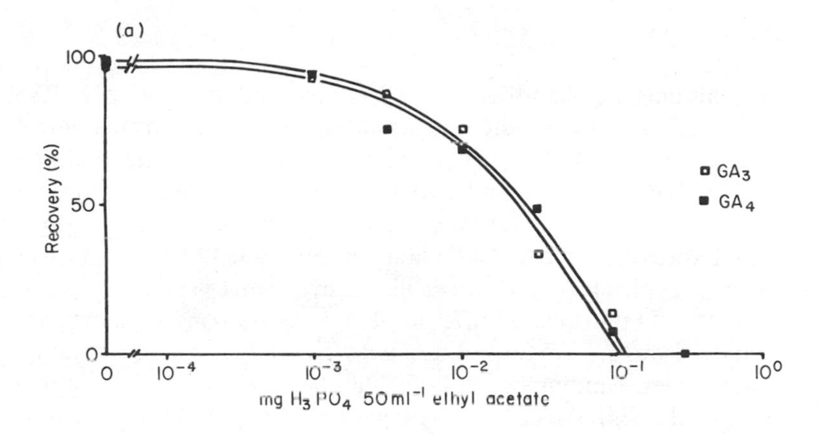

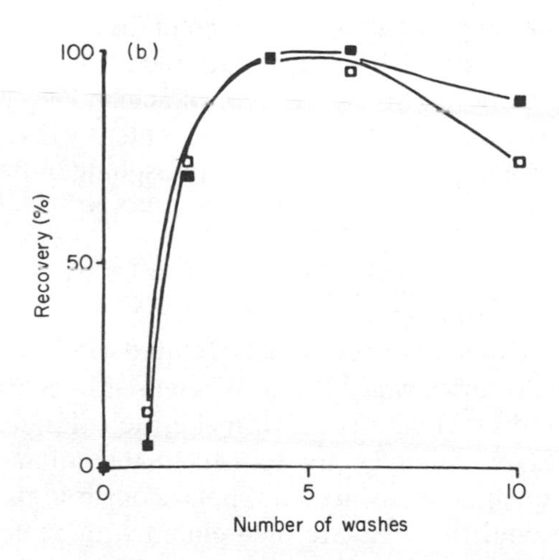

Fig. 7. (a). The effect of the presence of orthophosphoric acid on the recoveries of GA_3 and GA_4 after concentration of ethyl acetate solutions to dryness. 1 μM solutions of the GAs in 50 ml water-saturated ethyl acetate, containing different quantities of orthophosphoric acid as indicated, were concentrated at 30°C. Recoveries were determined by gas chromatography. (b). The effect of washing on the recoveries of GA_3 and GA_4. The ethyl acetate-soluble acid fraction (Fig. 6) was washed with 5 vol % water that had been adjusted to pH 2.5 with hydrochloric acid for the number of times indicated before concentrating to dryness at 30°C. Reproduced, with permission, from Rademacher (1978).

stable and result in altered chromatographic behaviour (Nutbeam and Briggs, 1982) as well as loss of biological activity (Corcoran *et al.*, 1972). The formation of these complexes displays some structure specificity both for the phenol and the GA. Certain phenolics can be effectively removed with insoluble polyvinylpyrrolidone (PVP).

(1) Polyvinylpyrrolidone (PVP)

Although binding of phenolics to PVP is favoured at acidic pH, PVP is generally used with the acidic extract in aqueous solution at neutral or slightly alkaline pH. At low pH, compounds with carboxylic acid groups may bind to PVP (Andersen and Sowers, 1968) so that GAs with more than one such group have large elution volumes (Glenn *et al.*, 1972; Sandberg *et al.*, 1981). However, even at pH 8.5, a significant reduction in sample weight is achieved. Typically the dried acidic ethyl acetate-soluble fraction is dissolved in 0.1 M phosphate buffer at pH 8.5, poured on to a short column containing 0.5–1.0 g PVP and eluted with about two column volumes of phosphate buffer. Binding of phenolics can be enhanced without affecting GA elution volumes if the column is prewashed with pH 3 buffer. In order to ensure reasonably fast flow rates, fine particles can be removed either by sieving or by slurrying and decanting. An alternative to columns is to make a slurry of PVP in an aqueous solution of the extract buffered at pH 8.0–8.5 and to separate the PVP by filtration. This is repeated with clean PVP until no further pigment binds to the PVP. This step can be conveniently inserted into the scheme in Fig. 6 after back-partition into 0.1 M phosphate buffer at pH 8.5.

(2) Charcoal–celite

GAs adsorbed on charcoal from aqueous solution can be eluted with acetone in order of decreasing polarity (Durley *et al.*, 1971). When used in columns the charcoal is generally mixed (1:2) with celite to make the column less retentive and increase flow rates. Samples are applied to the column in aqueous solution at low pH or in aqueous acetone, polar constituents are eluted with the same solvent and the GAs are then eluted with acetone. Batches of charcoal vary considerably in their activity, and each batch should be tested before use. The size of the column required will depend on the activity of the charcoal and the weight of the extract.

(3) Reversed-phase cartridge

A Sep-Pak C_{18} reversed-phase cartridge (Waters Associates, Inc.) consists of a short, disposable cartridge packed with silica gel to which octadecylsilane (ODS) is chemically bonded to form a stationary phase. The cartridges allow quick, convenient reversed-phase separations of small amounts of extract. However, due to their limited size they are easily overloaded and should be used late in the purification. In a typical procedure the Sep-Pak is first primed by pushing through it 5 ml methanol followed by 5 ml 5% acetic acid with a syringe. The extract is applied in 0.1 M phosphate buffer, pH 2.5, in a similar manner. The cartridge is then washed with 5 ml 5% acetic acid followed by 5 ml water. Finally GAs are eluted with 80% methanol–water. The elution profiles for GA_3 and GA_9 obtained using the foregoing procedure are illustrated in Fig. 8. It should be recognized that greater purification might be obtained by "fine-tuning" the eluting solvents. Thus more of the polar contaminants could be removed by pre-eluting with, for example, 20% methanol–water, although there is a risk of losing the more polar GAs. In fact the composition of the eluting solvents is best determined empirically to provide optimal recoveries and purification for the particular GAs to be analysed.

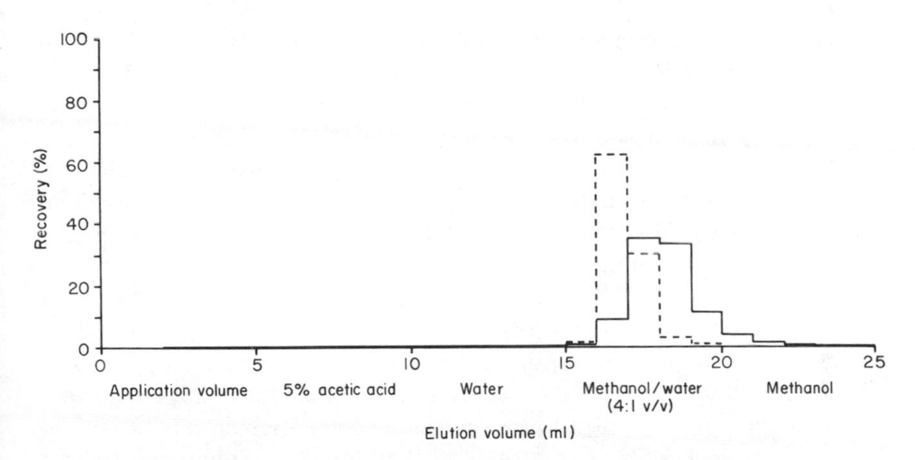

Fig. 8. Elution profiles of [^{14}C]GA_3 (---) and [^{3}H]GA_9 (——) from a Sep-Pak C_{18} cartridge. The sample was applied in 0.1 M phosphate buffer at pH 2.5 (5 ml) and eluted as described in the text, except for a final rinse with methanol (5 ml).

(4) Anion-exchange

Anion-exchange chromatography is potentially a very useful method that exploits the similarity in acid strengths of the GAs. Crozier and Durley (1983) described an adaptation of the DEAE-Sephadex A-25 column used by Gräbner *et al.* (1976) for GA-group separation. The extract is loaded in 0.2 M acetic acid–methanol (1:1, v/v) on to a column of the Sephadex in the acetate form and the column is eluted with four void volumes of the same solvent. GAs are then eluted with two void volumes of 2 M acetic acid/methanol (1:1, v/v).

The use of Dowex 1X1-100 in the formate form has been described (Browning and Saunders, 1977; Sponsel and MacMillan, 1978). The extract was applied to a column of the resin in aqueous solution at pH 8.0. After washing the column with water at pH 8.0 the GAs were eluted with ethanol–1.0 M formic acid (4:1, v/v). The Dowex 1X1-100 column has a high sample capacity and, if used on a crude extract, can give a substantial initial purification. There seems no reason why combinations of ion-exchange resins and small concentrator columns of silica-bonded ODS (Koshioka *et al.*, 1983b) or Amberlite (Andersson and Andersson, 1982) may not replace solvent–solvent partition, which is both inconvenient and time consuming.

(5) Gel filtration

The purification of extracts by passing buffered aqueous solutions through columns of Sephadex G-10 (Crozier *et al.*, 1969) or Sephadex LH-20 (Sandberg *et al.*, 1981) has been described. In both cases it appears that the chromatography is based on adsorption phenomena rather than steric exclusion effects. Sandberg *et al.* (1981) combined insoluble PVP with Sephadex LH-20 in a single 10 mm i.d. column (25 cm PVP and 20 cm Sephadex LH-20). GAs were eluted with 0.1 M sodium phosphate–citrate buffer at pH 8.0 in a narrow band.

Genuine size-exclusion purification was described by Reeve and Crozier (1976) who used long (100 × 2.5 cm) columns of Bio-Beads SX-4 (Bio-Rad Laboratories) eluted with tetrahydrofuran (THF). As expected, the GAs were eluted without appreciable separation. High-performance gel permeation systems have recently been described by Crozier *et al.* (1980) and Yamaguchi *et al.* (1982). The former workers used a μSpherogel support and THF or 0.1 M acetic acid in THF as solvent systems. Yamaguchi *et al.* used Shodex A-801 and also THF as the eluting solvent. In both cases the GAs (and GA conjugates) were eluted as a tight band in high yield.

D. Gibberellin separation methods

The final purification steps usually involve chromatographic procedures that result in some separation of individual GAs. This is inevitable because of the broad range of GA polarities. The most commonly used techniques are still classical chromatography procedures, which are relatively inexpensive and simple to execute. However, there is an increasing number of reports of the use of high-performance liquid chromatography (HPLC), which, because of its speed, high resolving power and reproducibility, will almost certainly take over as the method of choice.

(1) *Classical chromatography procedures*

(a) Thin layer chromatography. Reports of the purification of GAs by paper chromatography, although dwindling, still appear in the literature occasionally. The method is slow, offers very poor resolution and has a low sample capacity. Its main advantage was that it could be used in conjunction with bioassay without the need for elution; seedlings could be grown directly on segments of paper cut from the chromatogram. Paper chromatography has now been largely superseded by thin-layer chromatography (TLC), which is probably still the most common chromatographic method used for GA separations. It offers better resolution than paper chromatography and, with commercially coated plates and well-equilibrated solvent systems, should give reproducible separations. Its main disadvantages are that the resolution is inferior to many more recently developed methods and recoveries can be very low. Suitable solvent systems for GAs have been published (Sembdner *et al*., 1962; Kagawa *et al*., 1963; MacMillan and Suter, 1963; Cavell *et al*., 1967). Systems based on silica gel adsorption usually contain acidic (acetic or formic acid) solvents, which ensure that the GA carboxylic acid groups are protonated. This results in greater mobility and sharper bands. Basic systems such as 2-propanol/water/25% ammonium hydroxide (10:1:1, v/v) cause ionization of the GAs and as a consequence only the monocarboxylic acids, such as C_{19}-GAs, are mobile. GAs are eluted from silica gel with water-saturated ethyl acetate or with acetone. When a basic solvent system has been used the GAs must be eluted with methanol, which under basic conditions dissolves silica only very slightly.

Heftmann and Saunders (1978) have described the separation of GA esters on silica gel impregnated with silver nitrate. Since silver ions bind to double bonds this system will separate GA pairs, such as $GA_{4/7}$ or $GA_{1/3}$, that differ only in the presence or absence of a double bond. The GAs must be esterified first since free acids form salts with the silver ions and are eluted very poorly. Heftmann and Saunders formed *p*-nitrobenzyl esters, which absorb strongly in the UV and could be used to quantify the GAs. However, this method is unlikely to have much utility in the quantitative analysis of endogenous GAs, because of the poor resolution of TLC and the non-specificity of the esterification reaction. Silver nitrate-impregnated silica gel developed with benzene has been used to separate the GA precursor, *ent*-kaur-16-ene, from its 15-ene isomer (Bearder *et al*., 1974).

(b) Adsorption column chromatography. Large quantities of extract are commonly separated on columns of silica gel eluted with solvent gradients (see for example Beale *et al*., 1984). The method provides high sample

capacity but low resolution. Since recoveries of small samples can be poor, the method is not suitable for routine quantitative analysis.

Charcoal–celite columns eluted with gradients of acetone–water can also be used to separate GAs when large amounts of extract have to be purified (Durley *et al.*, 1972). In this case the more polar GAs are eluted first. However, such columns are notoriously slow and give unpredictable recoveries. Both silica gel and charcoal–celite columns have commonly been used in the complex purification schemes employed to obtain pure crystalline GAs from very large quantities of plant material (see for example Fukui *et al.*, 1977a). The advantages of these systems are that they are relatively inexpensive and can be easily increased in scale to handle very large extracts. However, with the improved sensitivity of modern analytical equipment it is seldom necessary to extract large quantities of material, and techniques offering superior resolution, albeit with reduced sample capacities, are preferable.

(c) Partition chromatography. Liquid–liquid partition columns, in which one liquid phase, usually the more polar one, is stationary and the second liquid phase, the mobile phase, is used to elute the solutes, can achieve very impressive separations. Durley *et al.* (1972) adapted the method originally described by Powell and Tautvydas (1967) in which silica gel was used to support the stationary phase. Durley *et al.* used 0.5 M formic acid as the stationary phase and a gradient of ethyl acetate in hexane saturated with stationary phase as the mobile phase. They obtained the most satisfactory separation of GAs when Woelm Silica Gel for Partition Chromatography was used as the support (see Table II). Other batches of silica gel were found to give less reproducible retention volumes because of the loss of stationary phase. However, Crozier and Durley (1983) pointed out that this drying out can be overcome if the solvents are degassed before use, and then the origin of the silica gel is much less critical.

The use of Sephadex supports for partition chromatography of GAs was introduced by Pitel *et al.* (1971) and Vining (1971). They used Sephadex G-25 to separate limited numbers of specific GAs. MacMillan and Wels (1973) developed a partition column using Sephadex LH-20 which gives excellent separation of a wide range of GAs. For example, the profile in Fig. 9 was achieved using the solvent mixture petroleum ether/ethyl acetate/acetic acid/methanol/water (100:80:5:40:7, v/v). The column was first equilibrated in the lower phase and then eluted with the upper phase until no more of the lower phase was co-eluted. The GA mixture was applied in the lower phase and eluted with the upper phase, which was maintained in equilibrium with the lower phase to avoid stripping it from the column. A column efficiency of 5500 theoretical plates was obtained for GA_3. The main

Table II. Elution of GAs from a Woelm silica gel partition column.

Fraction no.	Gibberellin	Fraction no.	Gibberellin
2	GA_9, GA_{12}	13	GA_1, GA_3
3	GA_{14}, GA_{24}, GA_{31},	14	GA_{19}
4	GA_4, GA_5, GA_6,	15	GA_2, GA_{19}
	GA_7, GA_{14}, GA_{15},	16	GA_2, GA_{13}, GA_{22}
	GA_{20}, GA_{25}, GA_{31}	17	GA_{18}, GA_{22}, GA_{26},
5	GA_{10}		GA_{29}
8	GA_{27}, GA_{34}	18	GA_{18}, GA_{26}, GA_{29}
9	GA_{27}, GA_{34}	19	GA_{17}
10	GA_{16}, GA_{27}, GA_{33},	20	GA_{23}
	GA_{34}	21	GA_{21}, GA_{23}
11	GA_{33}	23	GA_8, GA_{28}
12	GA_{30}	24	GA_8, GA_{28}

Reproduced from Durley *et al.* (1972).

Note: The silica gel column (20 × 1.3 cm) contained 20% water as the stationary phase. The mobile phase was a gradient of 65→50→100% ethyl acetate in hexane over 160 min. The solvents were saturated with 0.5 M formic acid. 20 ml fractions were collected.

GAs that appear in a single fraction were eluted at least 80% in that fraction.

disadvantage of this system is the low flow rate (50 ml h^{-1}), which results in very long analysis times. However, the 1450 × 15 mm column used by MacMillan and Wels has a high sample capacity and is suitable for separating fairly crude extracts. By adjusting the solvent composition the separation range can be altered at will.

(2) Countercurrent chromatography

Liquid–liquid partition chromatography is the basis for some recently developed techniques that have been used to separate plant hormones including GAs. These methods are adaptations of the countercurrent distribution that was extensively used in earlier GA separations (Crozier *et al.*, 1969).

In droplet countercurrent chromatography (DCCC) a series of columns is filled with one of the two phases (lighter or heavier), which acts as the stationary phase. The other phase (the mobile phase) is passed through the stationary phase in a series of droplets. When the lighter phase is the mobile phase (normal-phase chromatography) it is delivered to the bottom of the first tube, allowed to ascend to the top and then delivered to the bottom of the next tube and so on. When the heavier phase is mobile (reversed-phase chromatography) it is applied to the top of the tubes. Separations are achieved according to the partition coefficients of the solutes between the two phases. The application of a commercial DCCC system to the separation of GAs has been described (Bearder and MacMillan, 1980). The choice

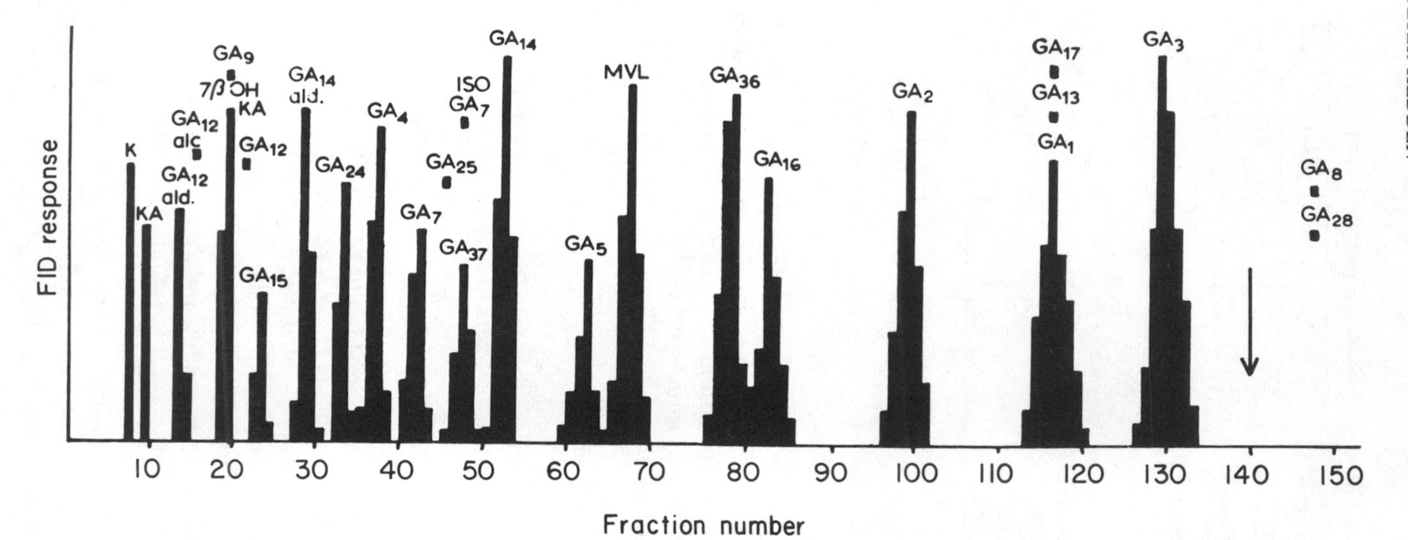

Fig. 9. Elution profile of GAs and related compounds from a Sephadex LH-20 partition column (147 × 1.5 cm). Stationary phase is the lower phase of the solvent mixture petroleum ether/ethyl acetate/acetic acid/methanol/water (100:80:5:40:7, v/v). The mobile phase is the upper phase of the above mixture and run at a flow rate of 50 ml h^{-1}. K, *ent*-kaurene; KA, *ent*-kaurenoic acid; GA_{12} ald., GA_{12}-7-aldehyde; 7βOH KA, *ent*-7α-hydroxykaurenoic acid; GA_{14} ald., GA_{14}-7-aldehyde; MVL, mevalonic lactone. Reproduced, with permission, from MacMillan and Wels (1973).

of solvents is limited by the requirement for the mobile phase to form droplets of suitable size in the stationary phase. This restricts particularly the choice of solvents for separating non-polar compounds. Although the resolution is low (about two theoretical plates per tube) large amounts of sample can be separated using small volumes of solvent, and recoveries are very high.

An alternative form of countercurrent chromatography was described by Mandava and Ito (1982). The authors used a toroidal coil planet centrifuge in which the stationary phase was loaded into 0.55 mm i.d. PTFE tubing wound into a coiled helix. The mobile phase was pumped through the tubing at 2.4 ml h^{-1} while the coil was centrifuged at 450–500 rpm. Although good resolution was achieved (2000–6000 theoretical plates), the analysis times were extremely long. GA_4 and GA_7 were resolved using ethyl acetate/methanol/0.5 M phosphate buffer (3:1:2, v/v) as the solvent system, but almost 10 h were required for the separation.

(3) High-performance liquid chromatography

High-performance liquid chromatography as it is now commonly called, although earlier reports refer to high-pressure liquid chromatography, is now taking over from other chromatographic methods, especially for the final stages of sample purification. The main advantages of HPLC over the other separation methods described so far are high performance (of the order of 500 theoretical plates per cm for GAs), highly reproducible retention volumes and short analysis times. In general terms the method consists of pumping the extract in liquid solution through a modified microparticulate support packed densely into a column. The support is usually 5–10 μm particles of silica gel. Smaller particle sizes allow denser packing and therefore higher column efficiencies. The particle shape can be irregular (i.e. Partisil) or spherical (i.e. Hypersil), the latter allowing more regular packing and generally better column efficiencies. Although untreated silica gel can be used for adsorption chromatography, it is more common to chemically bond a stationary phase to the silica and use partition chromatography as the mode of separation. For normal-phase chromatography a polar functional group is bonded to the support via a short carbon chain, whilst for the reversed-phase separations hydrocarbon chains such as C_8 (RP-8) or C_{18} (RP-18 or ODS) are bonded to the silica. Ion-exchange phases in which tertiary or quaternary ammonium groups (anion exchange) or sulphonic acid groups (cation exchange) are attached to the support are also available.

Samples are usually applied to the column dissolved in the mobile phase. A popular method is to inject the solution into a loop, from which it is flushed onto the column by the mobile phase. A single pump is required for isocratic elutions. Two pumps are generally used to produce solvent gradients although satisfactory results can also be obtained when solvents are directed to a single pump via a proportioning valve. Solvents should be of high purity to ensure reproducible retention times and should be degassed to prevent the formation of air bubbles after mixing two solvents in a gradient. The most effective (and expensive) method of degassing is to maintain a small flow of helium through the solvents. In order to protect the column it is common practice to pass the sample through a millipore filter before injection. The sample capacity of a column depends on the degree of loading of the stationary phase on the support and on the size of the column. Preparative columns are typically 5–10 mm i.d., although much larger columns, requiring higher flow rates, can be used for very large samples. Sample capacity may also be limited by the solubility of the extract in the mobile phase.

(a) Normal-phase chromatography. Reeve *et al*. (1976) described the first use of HPLC for GA separations. In their system, based on the silica gel partition column of Powell and Tautvydas, a 0.5 M formic acid stationary phase was coated onto a Partisil 20 silica gel support. A gradient of hexane–ethyl acetate was used as the mobile phase. The high loading of the stationary phase (40%) resulted in large sample capacities. Furthermore, the flow rate and separation efficiency were an order of magnitude better than in the original partition column. The disadvantage of the system was that the stationary phase was gradually stripped off the support unless precautions were taken to ensure that the mobile phase was always saturated with the stationary phase. Chemically-bonded phases overcome this problem. Morris and Zaerr (1978) separated GA 4-bromophenacyl esters on cyanopropyl silica eluted by a linear gradient from hexane/chloroform (90:10, v/v) to chloroform/2-propanol (100:1, v/v). More recently Crozier *et al*. (1982) separated GA methoxycoumaryl esters on CPS-Hypersil (also a CN-based phase) using isocratic elution with 3% ethanol in dichloromethane–hexane. The elution sequence is in order of increasing polarity. GA esters were used in both cases to provide a sensitive method of detection. This will be discussed in the following section. There have so far been relatively few applications of normal-phase HPLC to GA analysis in plant extracts.

(b) Adsorption chromatography. Little use has been made of adsorption HPLC for GA analyses. Recoveries are expected to be inferior to partition methods, and, if insufficient care is taken to establish and maintain solvent equilibrium, band broadening may occur. Lin and Heftmann (1981) published the separation of free GAs on silica (Zorbax BP-Sil) eluted

isocratically with n-hexane/ethanol/acetic acid (93:7:0.5, v/v). Particularly good separations of pairs of GA *p*-nitrobenzyl esters differing in the presence or absence of a double bond were obtained on silver nitrate-impregnated silica eluted with hexane/methanol/dichloromethane (91:8:1, v/v) (Heftmann *et al.*, 1978). The separation of benzyl esters on silica has also been described (Reeve and Crozier, 1978).

(c) Reversed-phase chromatography. It is evident from the numerous reports of the application of reversed-phase HPLC to GA purification that this is the method of choice for many laboratories. In a recent example Jensen *et al.* (1986) published the retention times of a wide range of GAs, GA-glucosides and glucosyl esters on a Supelcosil LC 18 column eluted isocratically with different mixtures of methanol in aqueous phosphoric acid. Their data for GAs are listed in Table III. Koshioka *et al.* (1983a) have also published retention times for a similar, but less extensive range of compounds. They used μBondapak C_{18} columns eluted with gradients of methanol in 1% acetic acid. Other reports include papers by Barendse *et al.* (1980); Jones *et al.* (1980) and Lin and Heftmann (1981). Although the mechanism of separation in reversed-phase HPLC is apparently more complicated than simple partition between the non-polar stationary phase and the polar mobile phase, solutes are in general eluted in order of decreasing polarity. Since adsorption plays a minor part in the chromatographic process, recoveries are high, even for polar compounds such as GA-glucosides, which are otherwise difficult to chromatograph on a preparative scale.

The mobile phase is usually aqueous methanol or ethanol containing a small volume of acid to ensure protonation of the GAs. Phosphoric, acetic and formic acids have been used, but acetic acid has the advantage that it is relatively weak and can thus be concentrated without risk of destroying the GAs. Very little acetic acid is in fact required; $50\,\mu l\,l^{-1}$ is sufficient. Although isocratic systems can be devised to resolve limited numbers of GAs, solvent gradients are usually necessary to separate a broad range of GAs with differing polarities.

IV. ANALYTICAL METHODS

A. Qualitative analysis

The first step when analysing the GA content of a plant tissue is to establish the identities of the GAs that are present. Only then can suitable methods for their quantitative analysis be devised. The degree of difficulty of compound identification depends on whether or not the compounds have been characterized previously, although theoretically in both cases the

Table III. Reversed-phase HPLC retention properties of gibberellins (Jensen *et al.* 1986). Gibberellins analysed on a 250 × 4.6 mm i.d. 5 μm Supelcosil LC 18 column eluted isocratically at 1 ml min^{-1} with methanol in aqueous phosphoric acid at pH 3.0. Detection with an UV absorbance monitor at 208 nm. Data expressed as retention times (mins).

	Methanol (%)												
	20	25	30	35	40	45	50	55	60	65	70	75	80
GA$_8$	11.5	8.2											
GA$_{29}$	14.8	10.3	8.2										
GA$_{39}$	18.0	11.5	8.3										
GA$_{33}$	17.8	11.9	9.0										
GA$_{30}$	21.7	13.1	9.6										
GA$_{23}$	26.3	16.0	11.3										
GA$_{28}$		18.4	12.2	8.0									
GA$_{38}$		19.1	12.4	8.2									
GA$_{41}$		21.7	13.7	9.1									
GA$_{26}$		22.3	14.0	9.1									
GA$_3$		23.0	14.0	9.2									
GA$_1$		26.5	16.5	10.8									
GA$_6$			24.6	15.2	10.1								
GA$_{18}$			24.8	15.4	10.0								
GA$_{35}$			24.2	15.8	10.5								
GA$_{22}$			33.7	19.8	12.1	8.1							
GA$_{21}$				22.0	14.7	9.7							
GA$_{31}$				28.7	16.4	10.6							
GA$_5$					24.3	14.6	9.6						
GA$_{10}$					24.9	15.3	9.8						
GA$_{16}$					25.6	16.2	11.0	8.2					
GA$_{20}$					28.0	17.2	11.2	8.4					
GA$_{27}$					31.2	19.3	11.7	9.1					
GA$_{47}$					30.5	19.5	12.6	9.4					
GA$_{36}$						22.1	13.8	10.1	8.1				
GA$_{13}$						22.6	14.3	10.0	8.0				
GA$_{40}$						23.0	14.3	10.0	8.4				
GA$_{44}$						25.7	15.2	11.0					
GA$_{19}$						32.9	19.0	13.4	9.5				
GA$_{34}$						32.0	21.0	14.9	10.7				
GA$_{51}$							22.8	15.3	10.8				
GA$_{17}$							24.8	16.2	11.0				
GA$_{37}$							24.2	16.8	11.6				
GA$_7$							30.0	20.6	13.3	8.4			
GA$_4$							36.8	25.0	15.3	9.7			
GA$_{14}$								38.0	21.6	12.6	8.1		
GA$_{53}$										13.9	9.5		
GA$_{24}$								42.2	23.6	13.8	10.0		
GA$_9$								43.0	24.6	15.0	10.4		
GA$_{15}$									26.7	17.6	11.1		
GA$_{12}$											25.5	17.5	11.5
GA$_{12}$-aldehyde											33.0	21.5	14.1

procedure is the same—sufficient information is accumulated about the unknown until a single structure can be assigned to it. However, in practice proving that two compounds are identical is much easier than elucidating an unknown structure. Thus the following section is divided into these two categories.

(1) Identification of known compounds

The most powerful chromatographic systems currently available still provide insufficient resolution to separate a particular compound from the millions that could be present in partially purified plant extracts. However, the combination of such a system with a discriminating detector, such as a mass spectrometer, will greatly enhance the chances of making a correct identification. In combination with an efficient chromatographic system such as capillary GC, mass spectrometry (MS) offers the best opportunity to identify small (ng) amounts of compounds in a purified extract, provided the compounds are previously known. The use of GC–MS for GA analysis has been pioneered by MacMillan and co-workers over the last 15 years (see for example MacMillan, 1972). The following discussion of its current use will cover GC and MS in turn (see also Hedden, 1986). It begins with a description of derivatization methods, which are required to increase GA volatility prior to GC analysis.

(a) Derivatization for gas chromatography and gas chromatography–mass spectrometry. The carboxylic acid groups of GAs are usually converted to methyl (Me) esters with diazomethane. The reagent is a yellow gas, which is prepared and stored in ether solution (see Schlenk and Gellerman, 1960 and Chapter 3, Section VIIF). This reagent is toxic and potentially explosive, so it should always be handled with extreme care in a fume cupboard. Samples to be methylated are dissolved in methanol and the ether solution of diazomethane added dropwise until no further evolution of nitrogen is observed and the solution remains yellow. After a few minutes, excess diazomethane is removed by evaporating the solution to dryness under nitrogen. Alternatively, when large volumes are involved, excess diazomethane is destroyed by adding acetic acid and the solution is taken to dryness in a rotary evaporator.

The methyl esters of GAs are considerably less polar than the free acids. After methylation, polar contaminants may be removed if the extract is dissolved in a relatively non-polar solvent such as dichloromethane and transferred to a clean vial. The sample may be analysed by GC at this stage, but it is more common to convert hydroxylated GAs to trimethylsilyl (TMS) ethers. This conversion decreases the polarities of the GA methyl esters still further, but, more importantly, it improves their mass spectral characteristics. The TMS ethers are more likely than the free alcohols to give intense molecular ions, and they give characteristic fragmentation patterns that may aid in structure elucidation. The free acids may also be converted to TMS ether esters, which are useful derivatives for preparative GC as the free GA is easily regenerated by hydrolysis in water.

Numerous reagents are available for trimethylsilylation. The most common in current use for GAs are BSTFA (bis-trimethyl-silyltrifluoroacetamide) and MSTFA (*N*-methyl-*O*-trimethylsilyltrifluoroacetamide), which is slightly more volatile. Small aliquots of extract in small vials or glass ampoules are thoroughly dried under nitrogen or in a desiccator. Ampoules can be prepared conveniently from Pasteur pipettes. It is important to ensure that the extract is dry since both the reagent and derivative are decomposed by moisture. The reagents, which also act as solvent for GC, are added to the dried extract, the vial is capped or the ampoule is sealed in a flame, and the solution is heated to about 90°C for 30 minutes. Alternatively, the extract can be dissolved in dry pyridine and the silylating reagent added subsequently. The ampoules should be resealed as soon as possible after opening if the remaining sample is to be retained.

Both BSTFA and MSTFA are highly reactive silylating reagents. They are also quite volatile and are therefore well separated from compounds of interest on the GC column if they are used as solvent. The reagents can also be removed under vacuum in a desiccator and replaced with a more volatile solvent, which should be aprotic and dry. The main disadvantage of these reagents is their relatively large molecular size. Thus certain vicinal diols are incompletely silylated. In such cases the more traditional Sweeley reagent (hexamethyldisilazane/trimethylsilyl chloride/pyridine, 3:1:9, v/v) is preferred, as it is for silylating the glucose moiety in GA-glucosides. GAs, such as GA_{26} or GA_{33}, which contain keto functions, may be converted by silylating reagents to TMS enol ethers. Usually this reaction is incomplete and a mixture of derivatives is produced.

Permethylation has been investigated as an alternative to trimethylsilylation (Rivier *et al.*, 1981). This technique is particularly useful for derivatizing GA-glucosyl ethers and esters, since the methyl ethers have much lower molecular weights than the corresponding TMS derivatives. The GAs are first esterified with diazomethane in the usual way and then the methyl ethers are formed by treatment with methyl iodide and sodium hydride in dimethylformamide solution.

Other, more specific derivatives have been prepared as a means of confirming structure. Vicinal diols have been converted to acetonides by reaction with acetone in the presence of acid (Gaskin and MacMillan, 1975) or to boronates with n-butylboronic acid (Graebe *et al.*, 1974).

(b) Gas chromatography. A major advance in GC has been the introduction of capillary columns, which are about 10 times more efficient than packed columns. Quartz silica wall-coated open tubular (WCOT) columns have been especially useful for GC–MS application since they are very easily installed and can be fed directly into the ion source of the mass spectrometer. This allows maximum transfer of components from the GC column to the mass spectrometer, which has been an important factor in achieving the increased sensitivity of modern GC–MS instruments. Quartz WCOT columns are now available with a broad range of stationary phases, including phases chemically bonded to the glass surface. Columns with chemically bonded phases produce less bleed, can be used at higher temperatures and have a longer life than normally coated columns. They can also be easily decontaminated by flushing with solvent without removing the stationary phase.

The gas chromatographic behaviours of GA Me esters, Me esters TMS ethers (MeTMS) and TMS ether esters on packed columns have been published extensively (see for example Cavell *et al.*, 1967; Schneider *et al.*, 1975; Crozier and Durley, 1983). However, lists of retention times are of limited value since it is extremely difficult to reproduce precise operating conditions from one instrument to another. Furthermore, columns can vary considerably in their retention characteristics. In any case it is dangerous to base identification on retention times alone, even if several columns with different polarity phases are used. However, in combination with a specific detector such as a mass spectrometer, retention time is a useful additional parameter that can aid in identification. Since the resolution offered by capillary columns is sufficient to separate most GAs, it is seldom necessary to use more than one column. A non-polar phase such as SE-30 or OV-1 (BP-1 is the bonded phase equivalent) is more stable and can be used at higher temperatures than more polar phases. Retention times on such phases depend primarily on molecular weight. Since differences in molecular weight are enhanced by the formation of TMS ethers, these derivatives are better resolved than the free alcohols. Although absolute retention times will vary from one column to another and also for the same column as it ages, relative retention times should remain fairly constant. Table IV lists retention indices for a number of GA MeTMS derivatives obtained on a BP-1 WCOT column using a linear temperature programme. The GAs were co-injected with a series of straight-chained saturated hydrocarbons. These can be conveniently prepared by dissolving parafilm in hexane or a suitable non-polar solvent (Gaskin *et al.*, 1971). The retention indices were read from the linear plot of hydrocarbon carbon number against retention time (Van den Dool and Kratz, 1963). The retention index (RI) is related to retention time by

$$\mathrm{RI} = 100i\,\frac{R_x - R_n}{R_{n+i} - R_n} + 100n$$

where R_x is the retention time of the sample, and R_n and R_{n+i} are the retention times of hydrocarbons with respectively n and $n + i$ carbon atoms.

Table IV. Retention indices (Kovats indices) for the MeTMS derivatives of certain GAs. GC conditions: Bonded OV-1 WCOT fused silica column (25 m × 0.2 mm i.d.), helium inlet pressure 13 psi, column maintained at 60°C for 1 min, then heated at 20°C min^{-1} to 240°C and then at 4°C min^{-1} to 300°C.

GA	RI	GA	RI	GA	RI
GA_1	2676	GA_{18}	2643	GA_{35}	2651
GA_2	2767	GA_{19}	2612	GA_{36}	2623
GA_3	2714	GA_{20}	2512	GA_{37}	2791
GA_4	2533	GA_{21}	2713	GA_{38}	2974
GA_5	2504	GA_{22}	2701	GA_{39}	2778
GA_6	2598	GA_{23}	2744	GA_{40}	2552
GA_7	2548	GA_{24}	2489	GA_{41}	2820
GA_8	2818	GA_{26} diTMS	2853	GA_{43}	2725
GA_9	2386	GA_{26} triTMS	2792	GA_{44}	2816
GA_{10}	2601	GA_{27}	2922	GA_{45}	2521
GA_{12}	2395	GA_{28}	2722	GA_{47}	2638
GA_{12}-aldehyde	2408	GA_{29}	2680	GA_{51}	2538
GA_{13}	2606	GA_{30}	2647	GA_{53}	2524
GA_{14}	2516	GA_{31}	2580	GA_{54}	2626
GA_{15}	2651	GA_{33} diTMS	2676	GA_{63}	2725
GA_{16}	2648	GA_{33} triTMS	2785		
GA_{17}	2598	GA_{34}	2671		

The sample capacity of capillary columns is much lower than that of packed columns. As little as 0.1 μg of a component can overload a 0.2 mm i.d. WCOT column which has a *ca.* 0.3 μm stationary phase film. Therefore samples for capillary GC analysis require more extensive purification than those used for chromatography on packed columns. There is some compensation from the greater sensitivity of capillary GC, which requires less sample to be injected. Quartz silica WCOT columns of 0.3 mm i.d. and a 0.5–1.0 μm stationary phase film are now available and for GC–MS they probably provide the best compromise between sample capacity and flow rates.

Sample introduction for capillary columns is critical since for efficient chromatography only small solvent volumes (≈0.1 μl) can usually be allowed to pass through the column. However, the extremely low concentrations of GAs in most samples dictate that larger volumes must be injected and that the sample cannot be split at the injection port. Larger solvent volumes can be injected without impaired chromatography using the Grob splitless injection technique (Grob and Grob, 1974). In this method 1–2 μl of sample solution are injected without splitting and with the column temperature at least 20°C below the boiling point of the solvent. After $\approx$30 s the split on the injection port is opened so that any solvent remaining in the injection port is swept out. The column oven is then heated rapidly to the programme starting temperature and the chromatography allowed to proceed as normal. For quantitative work the Grob splitless injector can sometimes be unsatisfactory due to unreproducible loading of the sample onto the column. A more satisfactory technique may be on-column injection, in which the sample is injected via a syringe with a long, narrow needle directly into the column.

(c) Gas chromatography detectors. Unlike most other classes of plant growth substance, GAs contain no special features which facilitate specific physicochemical detection. GAs are usually detected with a flame ionization detector (FID), which responds to all combustible compounds in an extract and is therefore highly non-specific. Thus, GC–FID is not a useful technique for the identification of unknown compounds.

GAs can be monitored with an electron-capture detector (ECD) provided they are derivatized with an electron-capturing group. Thus GA methyl esters have been converted to heptafluorobutyl or trifluoroacetyl derivatives by reaction with the corresponding imidazoles (Seeley and Power, 1974; Küllertz *et al.*, 1978). Although the technique greatly enhances the sensitivity of detection compared with FID it gives little improvement in specificity since all free alcohols will be derivatized.

Techniques for detecting radioactive compounds after GC separation will be discussed in Section VI.

(d) Mass spectrometry. In GC–MS the mass spectrometer serves as a highly versatile GC detector. Electron impact (EI) mass spectra are produced by bombarding molecules under vacuum (10^{-6}–10^{-7} torr) with high-energy electrons (usually 70 eV). Under these conditions GA MeTMS derivatives undergo extensive fragmentation, but intense molecular ions (the unfragmented charged molecule) are also quite common. Although both positive and negative ions are produced, it is usual to measure positive ions in EI–MS. The mass spectra, which are a plot of the relative abundance of fragment ions against their mass/charge ratio (m/z), are reproducible and characteristic. Provided a reference compound or, failing that, reference mass spectrum is available for comparison, a reliable identification can be

made on the basis of a close match of the two spectra. Binks *et al.* (1969) have published EI mass spectra for the Me and MeTMS derivatives of GA_1–GA_{24}. A more complete catalogue of GA spectra is currently being prepared by P. Gaskin, J. MacMillan and K. A. G. MacNeil ("Mass Spectra of Gibberellins, Reference Spectra of Functional Derivatives", to be published by Wiley). Some representative mass spectra for the different structural types are shown in Fig. 10, and Table V lists abbreviated MS data. A more detailed analysis of GA fragmentation patterns is given in Section IV.A.2(a).

Chemical ionization (CI) MS is an alternative to EI that is particularly useful for compounds that give no molecular ions in EI. Ionization takes place indirectly in the presence of a large excess of a reagent gas. The gas is ionized by electron impact and transfers its charge to the sample by collision via a hydrogen or heavier fragment. Since the ionization process involves less energy than EI, less fragmentation occurs. The method is useful for determining molecular weight, but provides less structural information than EI. CI has been little used for GAs although this is likely to change as the technique becomes more routine. Crozier *et al.* (1982) used negative ion CI in methane for the analysis of GA methoxycoumaryl esters. These derivatives, prepared for HPLC determination, were not sufficiently volatile for GC and were introduced into the mass spectrometer by direct probe.

Mass spectrometers used for GC–MS are of two basic types, which differ in their mass analyser, the system by which charged fragments are separated according to their mass. In magnetic sector instruments, ions are separated in a magnetic field. Double-focusing instruments contain an electrostatic analyser in series with the magnetic analyser to increase the resolution. Double-focusing GC–MS systems are available with resolution up to 1 in 10 000, allowing the elemental composition of the ions to be determined. Such information is extremely useful in structure determination. A disadvantage of older magnetic sector spectrometers for GC–MS was their relatively slow mass scanning rates. They were consequently unsuitable for use with capillary GC columns, because their high separating power would be wasted. However, with the development of laminated magnets of which the field strength can be rapidly and precisely varied, scanning rates of less than one second per mass decade (50 to 500 amu) are possible.

In quadrupole mass spectrometers the ions are separated by passing between four parallel rods, which are charged with radiofrequency (rf) and direct-current (dc) voltages. Opposite rods are connected and are 180° out of phase with the other pair. Only ions of a particular mass pass the complete length of the rods for a particular voltage. The mass range is scanned by progressively changing the voltages on the rods but keeping the ratio of rf to dc voltage constant. Quadrupole instruments have become popular for

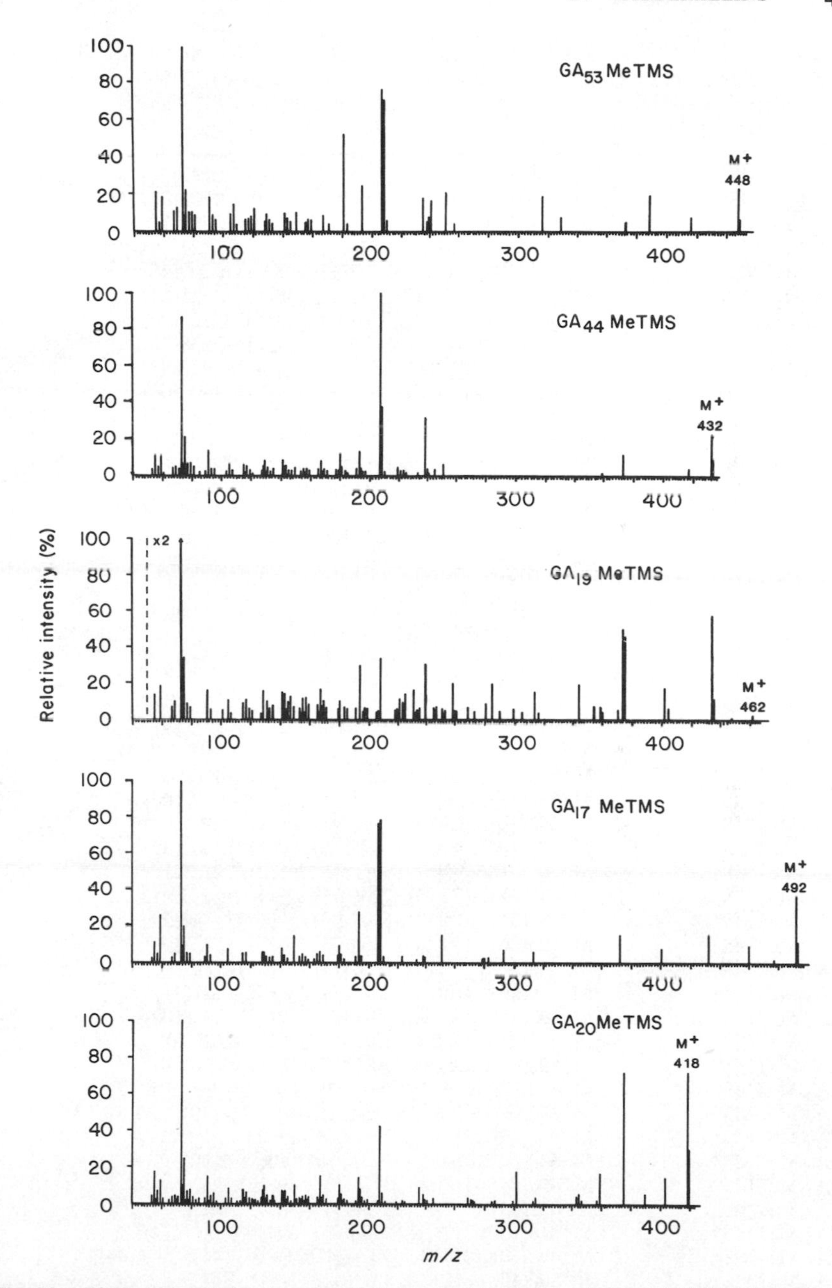

Fig. 10. Electron impact mass spectra of some representative GAs as MeTMS derivatives. The spectra were obtained by GC–MS using a VG 1212 mass spectrometer fitted with a Dani 3800 HR gas chromatograph and a 2015 data system. Mass spectrometer conditions: electron energy, 70 eV; emission current, 100 μA; source temperature, 200°C.

Table V. Five characteristsic ions in the EI–MS of GA_1–GA_{63} and GA_{68} as MeTMS derivatives. Ion intensities are normalized on the base peak except where this is at *m/z* 73 or 75, when the next most intense ion is used. Where the molecular ion (M^+) is very weak it is not included among the significant ions.

GA_1MeTMS[a]	M^+506(100%), 491(13), 448(20), 377(12), 313(17)
GA_2MeTMS[a]	M^+508(27), 493(15), 418(17), 289(19), 130(100)
GA_3MeTMS[a]	M^+504(100), 489(8), 370(9), 347(10), 208(45)
GA_4MeTMS[a]	M^+418(26), 289(70), 284(100), 225(82), 224(76)
GA_5MeTMS[a]	M^+416(100), 401(18), 357(13), 343(11), 299(25)
GA_6MeTMS[a]	M^+432(100), 417(18), 373(17), 302(68), 235(22)
GA_7MeTMS[a]	M^+416(18), 384(50), 356(66), 223(93), 222(100)
GA_8MeTMS[a]	M^+594(100), 448(25), 379(20), 375(15), 238(28)
GA_9Me[b]	M^+330(6), 298(100), 270(78), 243(43), 227(48)
GA_{10}MeTMS[a]	M^+420(40), 405(25), 363(13), 331(31), 130(100)
GA_{11}Me[a]	M^+344(100), 312(97), 284(42), 256(24), 240(26)
GA_{12}Me[b]	M^+360(2), 328(21), 300(100), 285(19), 240(31)
GA_{13}MeTMS[b]	M^+492(2), 477(9), 400(23), 310(27), 282(23)
GA_{14}MeTMS[b]	M^+448(5), 416(36), 388(23), 298(60), 287(60)
GA_{15}Me[a]	M^+344(23), 312(22), 298(13), 284(50), 239(100)
GA_{16}MeTMS[a]	M^+506(18), 416(20), 390(100), 360(43), 340(43)
GA_{17}MeTMS[a]	M^+492(73), 460(34), 432(37), 401(20), 373(39)
GA_{18}MeTMS[a]	M^+536(47), 521(11), 477(16), 319(29), 238(36)
GA_{19}MeTMS[b]	M^+462(10), 434(100), 402(41), 375(57), 374(59)
GA_{20}MeTMS[a]	M^+418(100), 403(14), 375(45), 359(12), 301(13)
GA_{21}MeTMS[a]	M^+462(100), 447(10), 430(18), 403(41), 345(13)
GA_{22}MeTMS[a]	M^+504(100), 489(28), 401(58), 387(20), 370(20)
GA_{23}MeTMS[a]	M^+550(21), 522(100), 463(24), 432(32), 373(38)
GA_{24}Me[b]	M^+374(4), 342(42), 314(100), 286(77), 226(89)
GA_{25}Me[a]	M^+404, 372(13), 312(82), 284(100), 253(9), 225(44)
GA_{26}MeTMS[a]	M^+520(100), 402(7), 255(8), 217(18), 147(33)
GA_{27}MeTMS[a]	M^+520(71), 430(11), 343(13), 223(25), 217(100)
GA_{28}MeTMS[a]	M^+580(31), 565(13), 371(14), 208(95), 207(100)
GA_{29}MeTMS[a]	M^+506(100), 491(11), 375(15), 303(17), 207(35)
GA_{30}MeTMS[a]	M^+504(9), 369(16), 280(12), 279(12), 221(32)
GA_{31}MeTMS[a]	M^+416(12), 282(45), 223(67), 222(100), 221(41)
GA_{32}MeTMS[c]	M^+680(51), 665(17), 590(100), 500(17), 339(39)
GA_{33}MeTMS	M^+520(26), 430(100), 383(36), 358(41), 237(50)
GA_{34}MeTMS[a]	M^+506(100), 288(9), 229(12), 217(22)
GA_{35}MeTMS[a]	M^+506(26), 416(34), 287(27), 282(42), 221(41)
GA_{36}MeTMS	M^+462(11), 430(58), 402(38), 312(47), 284(100)
GA_{37}MeTMS[a]	M^+432(9), 342(15), 310(25), 284(22), 282(15)
GA_{38}MeTMS[a]	M^+520(67), 505(8), 430(8), 238(22), 207(100)
GA_{39}MeTMS[a]	M^+580, 565(23), 488(22), 430(27), 398(36), 370(27)
GA_{40}MeTMS[a]	M^+418, 371(100), 343(83), 299(81), 284(80), 225(57)
GA_{41}MeTMS[a]	M^+582, 567(8), 423(6), 400(8), 283(13), 209(13)
GA_{42}MeTMS[a]	M^+538, 523(26), 416(22), 376(100), 287(92), 259(55)
GA_{43}MeTMS[a]	M^+580(9), 431(100), 371(15), 349(22), 217(58)
GA_{44}MeTMS[a]	M^+432(63), 417(12), 373(17), 238(41), 207(100)
GA_{45}MeTMS[a]	M^+418(100), 403(18), 358(23), 284(14), 225(14)
GA_{46}MeTMS[a]	M^+492, 460(41), 400(73), 342(74), 310(41), 282(100)
GA_{47}MeTMS[a]	M^+506(100), 459(9), 431(6), 313(11), 217(31)
GA_{48}MeTMS[a]	M^+594(58), 504(15), 419(11), 370(12), 191(42)
GA_{49}MeTMS[a]	M^+594(71), 504(14), 419(12), 370(13), 191(38)
GA_{50}MeTMS[a]	M^+594(22), 504(33), 460(12), 370(19), 309(22)
GA_{51}MeTMS[a]	M^+418, 386(25), 328(24), 284(100), 268(67), 225(91)

Table V—continued

GA_{52}MeTMS[a]	M⁺608(67), 518(10), 462(6), 342(9), 217(100)
GA_{53}MeTMS[a]	M⁺448(47), 389(25), 251(30), 241(18), 235(30)
GA_{54}MeTMS[a]	M⁺506(34), 416(46), 390(96), 375(43), 300(66)
GA_{55}MeTMS[a]	M⁺594(100), 553(12), 535(12), 448(22), 375(26)
GA_{56}MeTMS[a]	M⁺594(100), 522(15), 448(18), 379(11), 375(13)
GA_{57}MeTMS[d]	M⁺594(51), 579(7), 535(6), 448(100), 376(31)
GA_{58}MeTMS[e]	M⁺506(27), 416(51), 384(44), 356(45), 282(43)
GA_{59}MeTMS[f]	M⁺460(100), 401(34), 356(21), 343(32), 207(58)
GA_{60}MeTMS[g]	M⁺506(88), 491(16), 447(19), 375(100), 207(47)
GA_{61}MeTMS[g]	M⁺418(7), 359(41), 347(26), 296(96), 284(30)
GA_{62}MeTMS[g]	M⁺416, 401(4), 282(23), 223(100), 222(93), 180(36)
GA_{63}MeTMS	M⁺506(100), 446(40), 287(33), 282(28), 156(60)
GA_{68}MeTMS	M⁺504(82), 489(55), 370(77), 310(90), 280(90)

Source of information: [a]Crozier and Durley (1983), [b]Binks *et al.* (1969), [c]Bukovac *et al.* (1979), [d]Murofushi *et al.* (1980), [e]Gaskin *et al.* (1984), [f]Yokota and Takahashi (1981), [g]Kirkwood and MacMillan (1982). Otherwise mass spectra were obtained using either VG 1212 or Kratos MS80RFA instruments.

GC–MS because they are smaller and less expensive than the magnetic instruments and allow fast and reproducible scanning. They are particularly suitable for multiple ion monitoring (MIM), which will be discussed further in Section IV.B.1. On the other hand, they give low resolution mass analysis and have a limited mass range (up to 1000 amu) with a tendency to lose sensitivity at high mass values. However, these instruments are quite adequate for most routine analytical problems.

(e) Gas chromatography–mass spectrometry. The interface or connection between the GC column and ion source is one of the most important factors that determines overall sensitivity. Packed columns are coupled via a separator, which is necessary to remove most of the carrier gas. Glass jet separators are the most common in current use. Capillary columns are connected via a glass capillary tube or, in the case of quartz (fused silica) columns, can be coupled directly to the ion source. The carrier gas is usually helium, which because of its low molecular weight does not produce interfering ions. It is also easily removed in the jet separator as well as being the carrier gas of choice for capillary GC.

The introduction of dedicated computers for data handling has revolutionized GC–MS. The data systems not only process the mass spectral data, but can also control the instrument operation to the extent that complete automation is possible if not always desirable. Cyclic scanning of the mass range throughout a GC run generates hundreds of spectra, which can be handled practically only by computer. The stored data can be processed in several ways. A reconstructed total ion current (TIC) can be generated by summing the ion currents for each scan. The resulting trace is equivalent to an FID output. Limited current traces for ions of a particular *m/z* value can also be produced. These traces, which are known as mass chromatograms or cross scans, are useful as a means of picking out spectra of interest from the hundreds in a run. Figure 11 displays a series of mass chromatograms together with the TIC trace for ions in the mass spectrum of GA_{63} MeTMS, which is present in an extract of immature pear (*Pyrus communis*) seeds. This example illustrates the analytical power of mass spectrometry. Although GA_{63} is a minor component of the extract and is poorly distinguishable in the TIC trace, the mass chromatograms clearly indicate its presence at the expected retention time. Thus it is possible to search for suspected components by plotting mass chromatograms for characteristic ions. In this way compounds can be detected that are incompletely resolved from more abundant components and whose spectra are therefore contaminated by extraneous ions.

Although the presence of a compound in an extract may be indicated by mass chromatography, confirmation of identity requires comparison of a full spectrum with that from an authentic standard. The spectrum of interest can be recalled and viewed in normalized form, i.e. the ion intensities are plotted as a percentage of the most intense ion (the base peak). In practice it is often more desirable to normalize against an ion other than the base peak. TMS derivatives always produce an intense ion at *m/z* 73 that carries most of the charge. Other ions will be weak by comparison and are best normalized against a less intense ion. It is sometimes useful to intensify a part of the spectrum, usually the higher mass region, which may contain characteristic, but very weak, ions. Spectra may also be enhanced by subtracting extraneous ions. It is common practice to subtract a background spectrum, which contains ions due to GC column bleed, vacuum pump oil, etc., from the scan of interest. However, spectra may also be contaminated by ions due to components from which the compound of interest is incompletely resolved. Mixed spectra are quite common in complex extracts. If the compounds are slightly separated so that adjacent scans contain different proportions of the two spectra, clean spectra may be obtained by subtracting one scan from the other. Although this technique is useful when trying to interpret complicated mixed spectra, it should be practised with caution. The resulting spectra can be quite unrepresentative if too much is subtracted.

The reliability of an identification based on MS will depend on how well two spectra match. Reeve and Crozier (1980) used information theory in attempting an objective assessment of identification reliability. They calculated the number of binary digits (bits) of information necessary to distinguish a compound from all other possible components (estimated as 10^{42} with a molecular weight less than 1400) to be 140. Thus the degree of correspondence between the sample and reference spectra should exceed

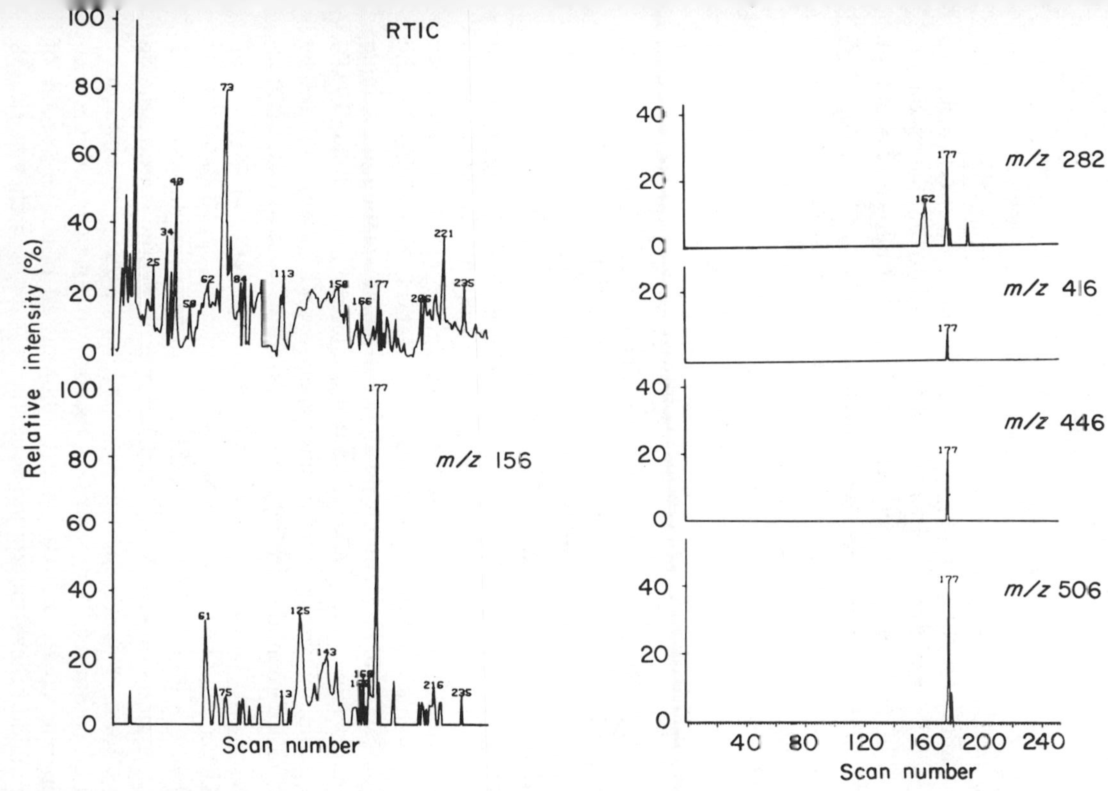

Fig. 11. Reconstructed total ion current (RTIC) trace for a methylated and trimethylsilylated extract of immature *Pyrus communis* seeds. Also shown are mass chromatograms for ions in the mass spectrum of GA_{63} MeTMS (see Fig. 15). The mass chromatograms are normalized to the *m/z* 156 trace. The traces were obtained using the GC–MS data system described for Fig. 10. Mass spectrometer conditions as for Fig. 10. Gas chromatograph conditions: sample injected on to a SE-30 WCOT glass capillary column (25 m × 0.2 mm i.d.) at 50°C with the split-closed. The split (50:1) opened after 0.5 min and after 1 min the oven heated ballistically to 240°C and then at 5°C min^{-1} to 300°C. Data collected from 240°C. Helium inlet pressure, 1.5 bar; injection port temperature, 200°C.

140 bits. Reeve and Crozier (1980) proposed that the degree of correspondence could be calculated by

$$I = \sum_{n=a}^{n=b} \frac{1.7 - \log|L_U - L_u|_n}{0.3} \text{ bits}$$

where $|L_U - L_u|_n$ is the absolute difference in relative intensities of the ions at $m/z\,n$ between the sample spectrum (L_U) and the reference spectrum (L_u), and a and b are the scanning limits. Reeve and Crozier assumed that each ion intensity is measured with a precision of ±3% so that the maximum information at each m/z value is 4.1 bits (minimum allowable value of $|L_U - L_u|_n$ is 3%). Furthermore, they estimated that only 25% of the ions can be used as a basis for comparison. This is because ions at low m/z values occur with high frequency and are common to many spectra. In practice, comparisons are based on relatively few ions and heavier ions are given more weight than lighter ones. Thus molecules which produce a low abundance of higher molecular-weight ions are more difficult to identify by mass spectrometry. The precision with which the relative intensity of ions can be measured will depend on the intensity of the spectrum. This depends on the amount of compound available and on the sensitivity of the instrument. Thus in many cases the information content at each m/z value will be less than that suggested by Reeve and Crozier.

In practice identifications based on mass spectra comparisons require a great deal of judgement, especially if only weak and/or highly contaminated sample spectra are available. Gas-chromatographic retention times are additional parameters that greatly aid identification. Knowledge of the retention time will put limits on the possible molecular weight of the sample and will similarly limit the number of possible molecules (MacMillan, 1984). If it is possible to compare both mass spectra and retention times an identification can be made with reasonable certainty. In some cases comparisons of retention times are essential. Certain GA epimers such as GA_{34} and GA_{47} or GA_{48} and GA_{49} give virtually identical mass spectra and can be distinguished using GC–MS only on the basis of retention times.

Most data systems for GC–MS contain libraries of spectra, stored either in abbreviated form or as full spectra, with which a sample spectrum can be compared by computer. Although GA spectra are not included it is relatively simple to create a library file containing all potential spectra of interest, provided these spectra are available. Library searches work well only with clean, relatively intense spectra, and aid in alerting the investigator to the possible identities of unknowns.

The sensitivity of a mass spectrometer can be enhanced up to 100-fold if only a limited number of ions are monitored. This technique, known as selective ion monitoring or multiple ion monitoring (MIM), is used mostly for quantitative analysis and will be discussed in more detail under that heading. The output is similar to a mass chromatogram. The technique is sometimes used for identification when only small amounts of sample are present (see for example Hedden *et al.*, 1982), but unless many ions are monitored simultaneously identifications are at best tentative.

(f) High-performance liquid chromatography. The lack of a selective means of detection has meant that HPLC is not well suited for the qualitative analysis of GAs. In common with all carboxylic acids, GAs absorb weakly in the UV and only below 220 nm. The sensitivity of the response can be increased by forming suitable derivatives. Table VI lists derivatives that have been prepared to enhance GA detection in HPLC. The most sensitive combination, *p*-methoxycoumaryl esters detected by spectrofluorimetry, enabled Crozier *et al.* (1982) to detect less than 1 pg of GA_3. However, since all carboxylic acids are similarly derivatized these methods do not increase the selectivity of detection to any great extent. This limitation was recognized by the authors who submitted HPLC fractions to further analysis by direct-probe MS. It is doubtful whether HPLC would provide sufficiently pure fractions from an extract for direct MS analysis unless very extensive prepurification was employed. There are few reports of HPLC being used directly for GA identification. There is, however, some potential for quantitative analysis, provided due care is taken to ensure the purity of the peak that is measured.

(2) Identification of new gibberellins

Although there are currently over 70 GAs for which the structures are known, each year a few more previously unidentified GAs are being discovered. The task of elucidating GA structures is more demanding than identifying known compounds and has been undertaken in few laboratories. It is, however, becoming progressively easier to characterize new structures, both because modern spectroscopic instruments are extremely sensitive and thus require very little sample and also because there are already many characterized GAs with which to compare spectroscopic properties. For these reasons methods have changed a great deal since the first elucidation of a GA structure, that of GA_3, by extensive chemical degradation almost 30 years ago (Grove, 1961).

Modern strategies are determined by the amount of compound available and the ease with which it can be isolated. If sufficient sample is at hand, chemical characterization of the compound can be attempted. This usually entails a thorough examination of its spectroscopic properties (infrared,

Table VI. Gibberellin derivatives used for detection in HPLC analysis.

Derivative	Derivatization procedure	Mode of detection (detection limit for mono ester)	HPLC method	Reference
Benzyl ester	GA + dimethylformamide dibenzylacetal in dioxane/70°C	UV-absorption λ_{max} 256 nm (300 ng)	adsorption	Reeve and Crozier (1978)
p-nitrobenzyl ester	GA + 0.1 M *O*-*p*-nitrobenzyl-*N*-*N'*-diisopropylisourea in dichloromethane/80°C	UV-absorption λ_{max} 265 nm (100 ng)	adsorption	Heftmann *et al.* (1978)
4-bromophenacyl ester	4-bromophenacyl bromide + 18-Crown-16 with GA K^+ salt 60°C	UV-absorption λ_{max} 256 nm (5 ng)	normal- and reversed-phase	Morris and Zaerr (1978)
p-methoxycoumaryl ester	4-bromomethyl-7-methoxycoumarin + 18-Crown-16 + K_2CO_3 in acetone	fluorescence $\lambda_{max}^{excitation}$ 320 nm $\lambda_{max}^{emission}$ 400 nm (1 pg)	normal- and reversed-phase	Crozier *et al.* (1982) Crozier and Durley (1983)

nuclear magnetic resonance and mass spectrometry) and some simple chemical conversions to derivatives or known structures. If, however, little sample is available and, as is often the case, the structure can be deduced from mass spectral information, the proposed structure can be synthesized and compared with the unknown.

(a) Gas chromatography–mass spectrometry. In almost all cases new GAs are first detected by GC–MS. Although a structure cannot be assigned to an unknown compound purely on the basis of its mass spectrum, EI mass spectra contain much structural information that helps in the subsequent investigation. Since high-resolution mass spectra are particularly informative, the use of high-resolution GC–MS is likely to be of great benefit in structure determination.

The following discussion of fragmentation patterns is based on MeTMS derivatives. The molecular ion is itself a guide to the structural type of the GA and the number of hydroxyl groups (see Table V). Although 13-hydroxylated GAs often give intense molecular ions, for many other GAs this ion is very weak. If the *m/z* value of the molecular ion is in doubt, there are characteristic losses that may help to establish its position. Losses of 90 amu (TMSOH) are common from TMS ethers. A primary TMS ether such as in GA_{22} loses 103 amu (CH_2OTMS). Losses of 31/32 (CH_3O/H) and 59/60 (CH_3OCO/H) due to cleavage or rearrangement of the methoxycarbonyl group are usually very prevalent, especially when two or three such groups are present. C-10 aldehydic GAs such as GA_{24} or GA_{19} show losses of 28/29 (OC/H). Combinations of these losses will be apparent in most mass spectra.

A prominent ion at *m/z* 129 with structure **3** (Fig. 12) is characteristic of GAs with a saturated A ring containing a single hydroxyl group at either the 1 or 3 positions. Ions at 207/208, accompanied by an ion at *m/z* 193, are characteristic of 13-hydroxylated GAs containing no other hydroxyl groups on the C or D rings. The fragment ion at *m/z* 208 presumably has structure **4**. However, some other GAs with a single hydroxyl group on ring C may also produce ions at *m/z* 207/8, e.g. GA_{30} and GA_{31}.

GAs such as GA_2 and GA_{10} with a 16-hydroxyl group give intense ions at *m/z* 130, presumably with structure **5**. 15-Hydroxylated GAs with no other hydroxyl group on the C or D rings give a characteristic ion at *m/z* 156. Vicinal diols as are typically found at the 2 and 3 positions in GAs such as GA_8 and GA_{34} produce intense ions at *m/z* 147 (**6**).

Loss of a fragment of mass 116 (**7**) is typical of 1,3 dihydroxylated GAs such as GA_{16} and GA_{54}. Monohydroxylated C_{19}-GAs such as GA_4, GA_{45} and GA_{51} produce ions at *m/z* 284 and 224/5 due to a combination of losses. The ion at *m/z* 284 $[M - 134]^+$ is due to losses of 90 and 44 (carbon dioxide

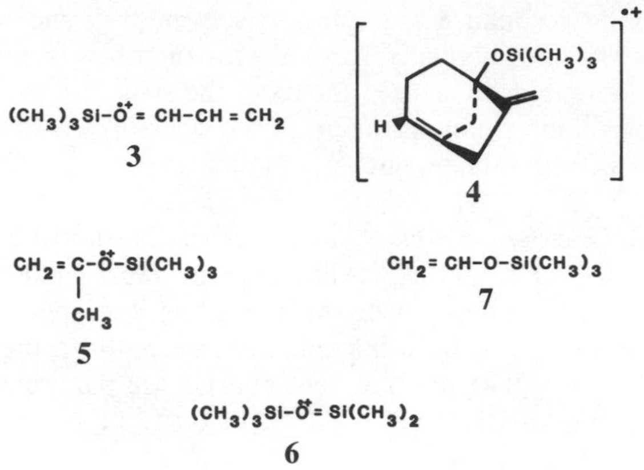

$(CH_3)_3Si\text{-}\overset{+}{\ddot{O}}=CH\text{-}CH=CH_2$

3

$[\ \ldots OSi(CH_3)_3,\ H\ ]^{\bullet +}$

4

$CH_2=C(CH_3)\text{-}\overset{+}{\ddot{O}}\text{-}Si(CH_3)_3$

5

$CH_2=CH\text{-}O\text{-}Si(CH_3)_3$

7

$(CH_3)_3Si\text{-}\overset{+}{\ddot{O}}=Si(CH_3)_2$

6

Fig. 12. Characteristic fragments produced in electron impact mass spectra of certain GA MeTMS derivatives.

from the γ-lactone), and further losses of 59/60 give rise to the ions at *m/z* 224/5.

Some of the losses discussed above are illustrated in Fig. 13, which outlines the fragmentation pattern of a dihydroxy C_{19}-GA MeTMS derivative.

(b) Nuclear magnetic resonance. Proton and ^{13}C nuclear magnetic resonance (NMR) spectroscopy and, to a lesser extent, infrared spectroscopy can give valuable information on the nature, number and position of functional groups. In particular, ^1H NMR can be used to deduce the configuration of hydroxyl groups, information that is not readily obtained from MS.

In NMR spectroscopy measurement is made of the absorption of energy supplied as radiofrequency (rf) electromagnetic radiation by certain atomic nuclei held in a magnetic field. This absorption corresponds to a change in direction of nuclear spin relative to the magnetic field direction. The resonance frequency $\nu = \gamma H/2\pi$, where H is the magnetic field strength and γ is a constant for a particular nucleus. Generally the rf radiation is held constant and the magnetic field strength is varied so that each nucleus in turn absorbs energy. The field strengths experienced by nuclei in a molecule vary slightly due to electronic distribution. The value, which is recorded relative to a standard (tetramethylsilane), is known as the chemical shift. Chemical shift values are characteristic for particular functional groups: also, spin–spin coupling with neighbouring nuclei resulting in a splitting of the signal, can give useful information on the nature and number of these nuclei.

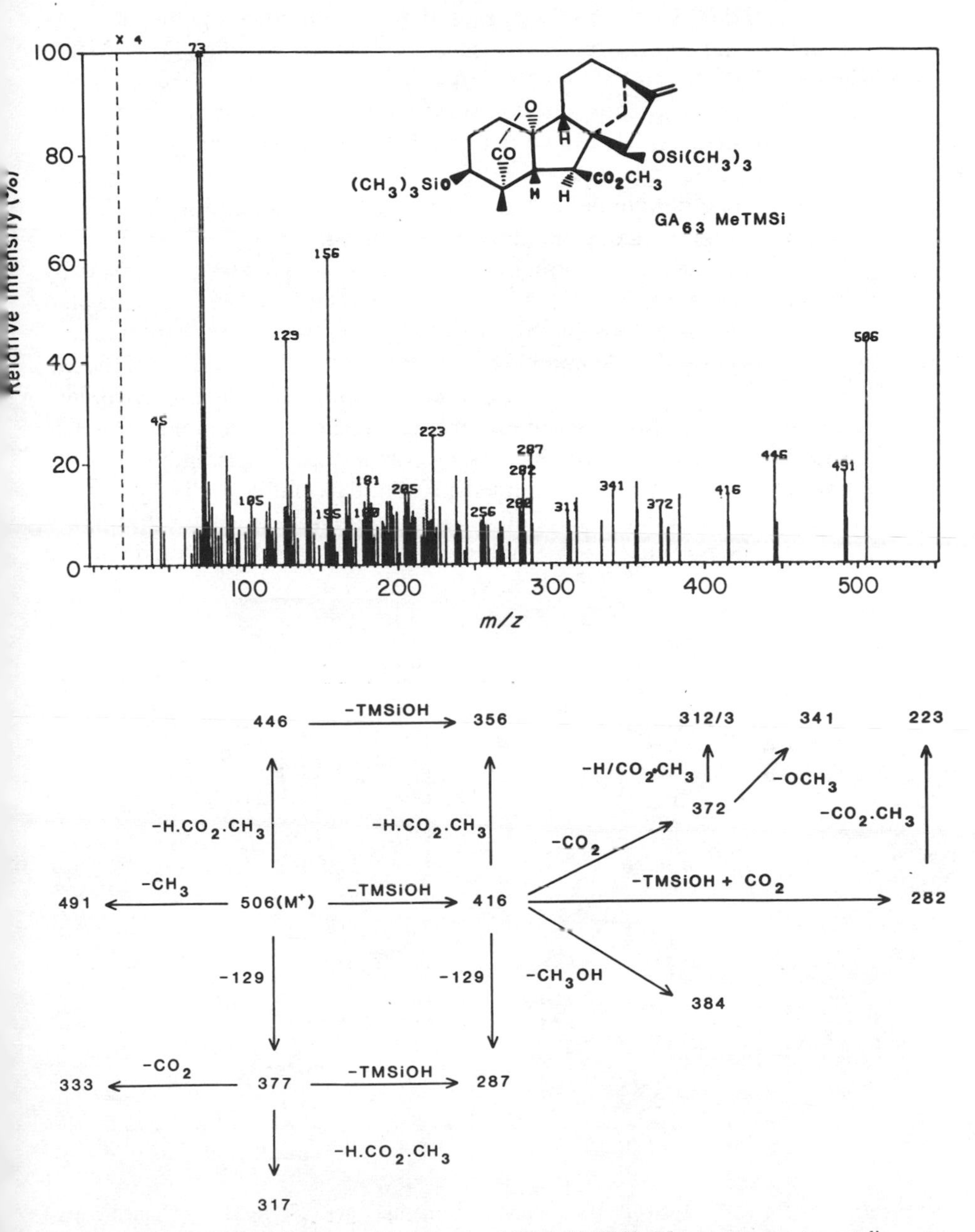

Fig 13. Principal losses in the mass spectrum of GA_{63} MeTMS. Mass spectrometer conditions as for Fig. 10.

C

Proton (^{1}H) NMR spectroscopy is used most commonly, but the introduction of Fourier transform NMR spectrometers has made ^{13}C NMR a valuable alternative. Proton-decoupled ^{13}C NMR spectra are generally simpler and easier to interpret than ^{1}H spectra. Table VII lists ^{13}C chemical shifts for some GAs. The data are taken from Yamaguchi *et al.* (1975a, 1975b) and Beale *et al.* (1984).

There are many examples of the use of NMR in the identification of GAs. Two cases are given from the more recent literature.

(i) GA_{59} was isolated from immature seeds of *Canavalia gladiata* by Yokota and Takahashi (1981). Mass spectral data on the Me ester indicated that the GA was similar structurally to GA_{21}, but was 2 amu lighter, i.e. it contained an extra double bond. The 100 MHz ^{1}H NMR spectrum of the Me ester in [^{2}H]trichloromethane is reproduced in Fig. 14. The proton assignments are added. These assignments were confirmed in a double resonance experiment in which proton decoupling was induced by irradiation at the resonance frequency of the coupled proton. This method allows coupled protons to be identified and confirms their proximity to each other in the molecule. The presence of two methoxycarbonyl signals confirmed that the

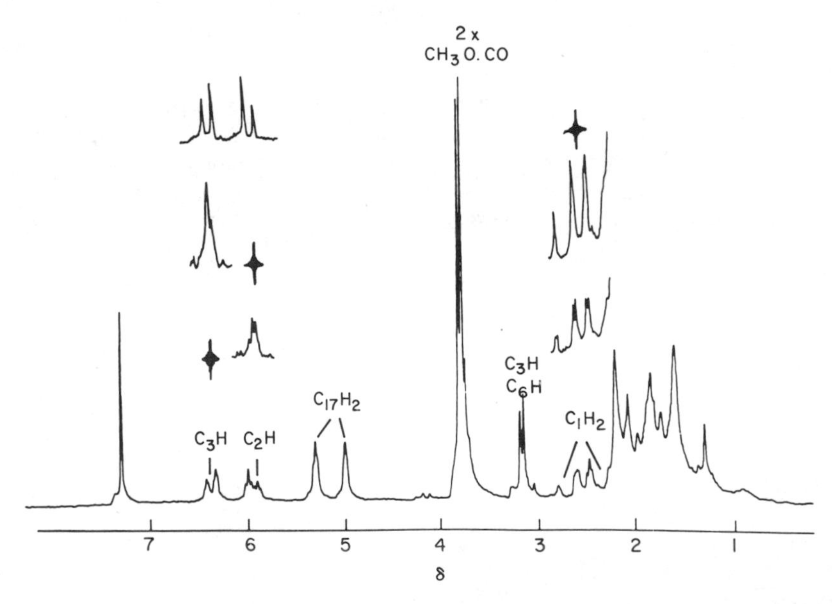

Fig. 14. 100 MHz ^{1}H NMR spectrum of GA_{59} methyl ester in [^{2}H]$CHCl_3$ obtained using a JEOL FX-100 spectrometer. The results of a double irradiation experiment are also shown. The second irradiation frequency is indicated as well as resulting changes in the splitting patterns. Reproduced, with permission, from Yokota and Takahashi (1981). The proton assignments are added.

GA was a dicarboxylic acid. It could be shown that the endocyclic olefin protons were coupled (and therefore adjacent) to methylene protons. Therefore the double bond could be situated only in ring A. Its position at 2,3 rather than 1,2 was suggested by the unusually low field absorption of one of these olefin protons (on C-3; $\delta 6.34$), which was assumed to be strongly deshielded by the C-4 methoxycarbonyl group. Finally the hydroxyl group was shown to be on C-13 from the downfield shift experienced by the exocyclic methylene protons when the solvent was changed from [^{2}H]trichloromethane to deuteropyridine. Such shifts are characteristic of 13-hydroxy GAs (Hanson, 1965).

(ii) GA_{58} was isolated from *Cucurbita maxima* endosperm after cellulase hydrolysis (Beale *et al.*, 1984). The position of the hydroxyl groups was indicated in the ^{13}C NMR spectrum by comparison of the chemical shifts with those of other GAs (Table VII). For example, the relatively large chemical shift of C-12 and the effect on the chemical shifts of C-16 and C-17 is typical of 12-hydroxylated GAs (Yamaguchi *et al.*, 1975a). The configuration of the 12-hydroxyl group was determined from the ^{1}H NMR spectrum of the Me ester. When the solvent was changed from [^{2}H]trichloromethane to deuteropyridine there was a large downfield shift of one of the C-17 protons. This shift, also seen with GA_{49}, but not GA_{48} (Fukui *et al.*, 1977b), suggested the α-configuration for the hydroxyl group. Additional evidence for the identity of GA_{58} came from GC–MS comparison of the products obtained from incubating *ent*-12β-hydroxykaurenoic acid with the fungus *Gibberella fujikuroi* (Gaskin *et al.*, 1984; Blechschmidt *et al.*, 1984).

(c) Partial synthesis. An alternative strategy, and the only one available when small quantities of the unknown compound are isolated, is to synthesize the suspected structure and compare its mass spectrum with that of the unknown. The new structure can be produced by partial chemical synthesis from another GA or *ent*-kaurenoid, or from incubations of suitable precursors with fungi.

The synthetic approach has been used in several recent characterizations. An example is illustrated in Fig. 15. The starting material is GA_3 or GA_7, which are available in large quantities from cultures of *G. fujikuroi*. The reader is referred to the original paper (Kirkwood and MacMillan, 1982) for details of the chemical methods. The reactions are relatively simple, but yield several GAs that are found in low concentration in their natural source.

Microbiological methods have also proved useful. The fungus *G. fujikuroi* converts certain analogues of GA-biosynthetic precursors to the corresponding GA analogues. The precursor analogues may occur naturally or be chemically synthesized. It is usually much simpler to introduce new func-

Table VII. ^{13}C NMR assignments for GAs (chemical shifts given as δ ppm).

	C-1	C-2	C-3	C-4	C-5	C-6	C-7	C-8	C-9	C-10
GA_{1}	28.2	29.2	70.0	55.5	52.5[a]	52.7[a]	175.2	49.8	53.5	93.9
GA_{3}	132.3	134.3	70.0	54.5	51.6	52.1	174.8	50.6	53.5	91.1
GA_{4}	28.2	29.2	70.0	55.5	51.6	52.8	175.2	51.5	54.0	94.1
GA_{5}	35.6	128.1[a]	133.1[a]	48.5	56.1	52.2	n.o.	50.7	53.5	91.9
GA_{7}	132.3	134.1	70.0	54.4	52.1	52.4	174.8	52.1	52.9	91.4
GA_{9}	30.9	19.8	34.6	49.1	58.2	52.8	174.8	51.3	54.1	93.0
GA_{13}	30.8	32.2	71.2	50.6	50.4	52.1	177.1	50.6	57.0	57.6
GA_{14}	28.4	34.6	71.0	49.5[a]	50.2	51.9	177.9[b]	50.0[a]	57.6	44.6
GA_{16}	71.8	39.7	70.4	54.8	51.1	53.0	174.9	52.1	54.0	96.2
GA_{17}	37.7	22.6	39.0	46.4	56.5	52.2	177.0[a]	48.7	56.9	57.7
GA_{18}	28.5	34.6	70.9	49.6	51.0	51.9	177.9[a]	48.6	57.2	44.7
GA_{23}	27.6	30.2	72.9	49.7	48.2	51.8	176.4[a]	52.8	56.7	48.4
GA_{28}	31.0	31.9	71.1	51.3	50.7	52.1	177.5[a]	48.9	56.6	57.3
GA_{30}	132.4	143.3	70.0	54.7	52.0	53.4	174.9	52.9	50.8	91.3
GA_{35}	29.4[a]	29.9[a]	69.9	55.6	52.2	52.9	175.2	51.9	61.0	94.6
GA_{36}	27.5	30.1	73.0	50.2	47.7	51.7	176.4[a]	50.2	56.7	49.5
GA_{37}	29.7	33.6	73.5	49.0	46.8	52.7	174.9[a]	49.9	56.1	41.9
GA_{38}	29.7	33.7	73.5	49.1	45.8	52.7	174.9[a]	47.8	55.9	41.6
GA_{39}	31.3	32.2[a]	71.6	51.2[b]	50.6[c]	52.2	177.2[d]	51.0[c]	54.1	58.4
GA_{40}*	39.1	64.6	45.3	46.3	58.7	52.6	175.3	51.7	54.2	92.8
GA_{58}†	27.9	28.8	69.6	55.6	50.2	50.7	175.3	51.6	52.1	94.3
	C-11	**C-12**	**C-13**	**C-14**	**C-15**	**C-16**	**C-17**	**C-18**	**C-19**	**C-20**
GA_{1}	18.0	39.9	77.9	46.3	44.0	159.2	106.5	15.6	179.0	—
GA_{3}	17.6	39.9	77.7	45.6	44.0	159.1	106.7	15.5	179.6	—
GA_{4}	16.5	31.8	39.4	37.4	45.0	157.7	107.2	15.5	179.0	—
GA_{5}	18.0	39.8	77.8	46.6	43.7	158.7	106.4	15.8	n.o.	—
GA_{7}	16.3	31.8	39.2	36.9	45.0	157.7	107.4	15.5	179.6	—
GA_{9}	16.4	31.7	39.3	37.2	44.8	157.6	107.1	17.7	178.8	—
GA_{13}	19.4	32.2	40.2	36.9	47.3	157.9	105.9	24.8	178.0[a]	178.5[a]
GA_{14}	17.3	32.6	40.6	39.8	47.1	157.3	105.7	25.4	180.6[b]	15.5
GA_{16}	18.7	32.3	39.1	37.0	45.2	158.2	107.4	14.9	179.3	—
GA_{17}	20.4	40.1	78.3	46.0	45.7	158.9	105.1	29.9	177.9[a]	178.4[a]
GA_{18}	18.9	40.8	78.3	49.1	46.2	158.7	105.3	25.1	180.7[a]	15.3
GA_{23}	19.7	40.2	78.2	45.9	45.9	159.1	105.4	22.6	177.0[a]	n.o.
GA_{28}	20.6	40.3	78.4	46.4	46.4	159.3	105.0	24.9	177.9	178.6[a]
GA_{30}	27.8	75.4	49.0	34.7	45.6	153.4	109.3	15.7	n.o.	—
GA_{35}	64.0	43.0	39.4	37.4	45.6	158.0	107.4	15.8	179.5	—
GA_{36}	18.6	32.2	39.9	37.2	46.9	157.6	106.2	22.6	n.o.[a]	n.o.
GA_{37}	16.1	31.9	39.9	36.7	45.4	157.9	106.4	21.1	175.8[a]	74.4
GA_{38}	17.2	40.0	78.3	45.8	45.1	159.6	105.4	21.2	175.6[a]	74.4
GA_{39}	31.8[a]	75.7	51.9[b]	33.4	48.4	154.7	107.7	25.3	178.0[d]	178.7[d]
GA_{40}*	16.6	31.8	39.4	37.3	45.0	157.8	107.3	18.4	179.8	—
GA_{58}†	27.4	75.1	52.6	34.6	45.3	153.0	109.5	15.4	179.7	—

Data reproduced from Yamaguchi *et al.* (1975a), except *(Yamaguchi *et al.*, 1975b) and †(Beale *et al.*, 1984). Samples were run in deuteropyridine at 25.15 MHz or 100 MHz†. *a–d* chemical shifts within a row may be reversed. n.o.: not observed due to short pulse interval.

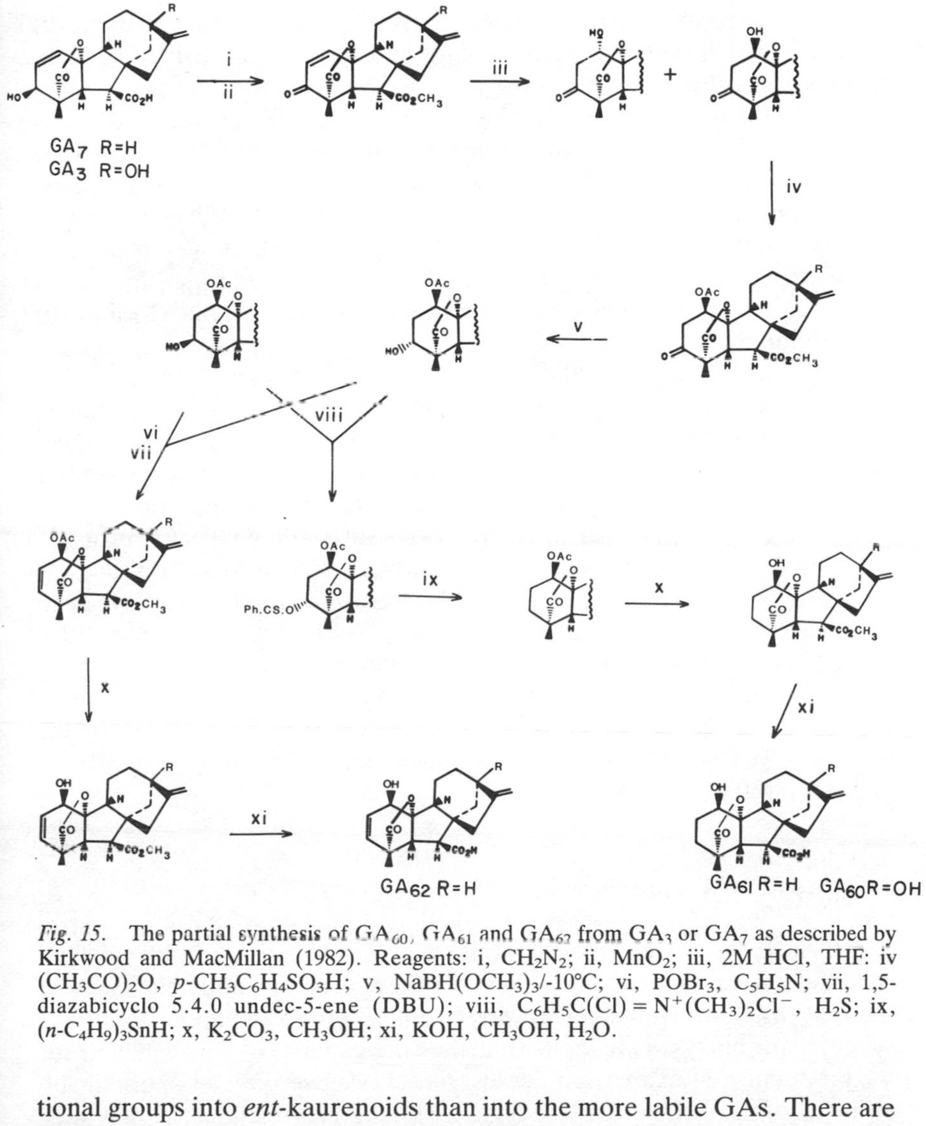

Fig. 15. The partial synthesis of GA_{60}, GA_{61} and GA_{62} from GA_3 or GA_7 as described by Kirkwood and MacMillan (1982). Reagents: i, CH_2N_2; ii, MnO_2; iii, 2M HCl, THF: iv $(CH_3CO)_2O$, *p*-$CH_3C_6H_4SO_3H$; v, $NaBH(OCH_3)_3$/-10°C; vi, $POBr_3$, C_5H_5N; vii, 1,5-diazabicyclo 5.4.0 undec-5-ene (DBU); viii, $C_6H_5C(Cl) = N^+(CH_3)_2Cl^-$, H_2S; ix, $(n\text{-}C_4H_9)_3SnH$; x, K_2CO_3, CH_3OH; xi, KOH, CH_3OH, H_2O.

tional groups into *ent*-kaurenoids than into the more labile GAs. There are numerous examples of the microbiological approach. Those from MacMillan's laboratory have involved the use of the GA-deficient mutant, B1-41a, which is blocked for GA biosynthesis just before *ent*-kaurenoic acid (Bearder *et al.*, 1974) and is therefore ideal for this purpose. However, if it is desirable to suppress endogenous GA synthesis there are several chemical inhibitors that can be used. GA_{45}, which occurs naturally in seeds of *Pyrus communis*, was identified as a product from an incubation of B1-41a with *ent*-15α-hydroxykaurenoic acid (Bearder *et al.*, 1975a). Wada *et al.* (1979) have reported the conversion of the same substrate to 15α-hydroxy GAs by *G. fujikuroi* cultures grown in the presence of an inhibitor (1-decylimidazole). A further example is illustrated in Fig. 16. A series of 12α-hydroxylated GAs isolated from *Cucurbita maxima* seeds was identified by comparison using GC–MS with products produced from synthetically derived *ent*-12β-hydroxykaurenoic acid (Lewis and MacMillan, 1980) with B1-41a (Blechschmidt *et al.*, 1984; Gaskin *et al.*, 1984). However, of these 12α-hydroxylated GAs, only GA_{58}, which was isolated in pure form, was considered to be sufficiently well characterized to warrant the allocation of an A number.

It may be apparent to the reader that the methods of identification described above depend on comparison with previously known compounds, either spectroscopically or through chemical interconversions. These compounds had been in turn related to others, and so on. Ultimately C_{19}-GAs are related to GA_3, whose structure was determined by chemical degradation and has been confirmed by X-ray diffraction (see for example Kutschabsky and Adam, 1983). C_{20}-GAs were related to *ent*-kaurene *via* 7β-hydroxykaurenolide (Cross *et al.*, 1963). Both classes of GAs have been chemically related by the conversion of GA_{13} to GA_4 (Murofushi *et al.*, 1976; Bearder *et al.*, 1979). The final confirmation of structure has now been accomplished by the total synthesis of several GAs. This remarkable achievement has been realized in several laboratories and has given new insights into GA chemistry (Fujita and Node, 1977; Corey *et al.*, 1978; Mander, 1980).

A. Quantitative analysis

The GAs have proved one of the most difficult classes of plant growth substances to measure quantitatively. Until relatively recently bioassay was the only method available. However, unless all the components of the system being analysed are known and the endogenous GAs are available for the production of calibration curves, bioassays give only an approximate estimate of GA content. Although bioassays are selective in that they respond to GAs and not other hormones, they exhibit varying degrees of response to individual GAs. Furthermore, the response is subject to interference by other components in the extract and lacks precision. Much more reliable data are obtained when GC–MS is used to quantify endogenous GAs. However, the recently-introduced immunoassay technique, provided it is used carefully with due regard to its limitations, may provide a viable alternative.

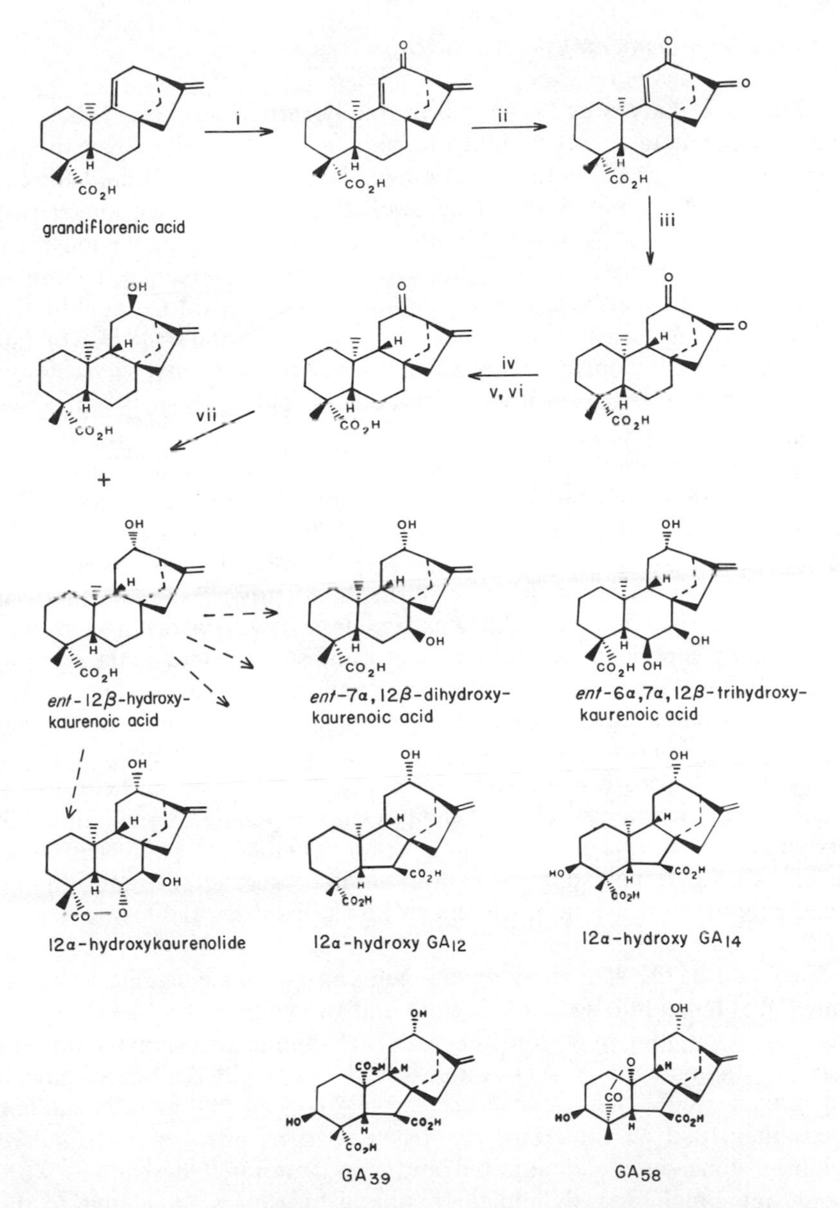

Fig. 16. Preparation of 12α-hydroxy GAs and *ent*-kaurenoid compounds. A mixture of *ent*-12β- and *ent*-12α-hydroxykaurenoic acids were prepared chemically from grandiflorenic acid (Lewis and MacMillan, 1980). Reagents: i, $C_4H_9CrO_4$; ii, OsO_4–$NaIO_4$; iii, 10% Pd–$CaCO_3$; iv, $(CH_3)_3SiCl$, $((CH_3)_3Si)_2NH$, C_5H_5N; v, $(C_6H_5)_3P = CH_2$; vi, HCl, vii, $NaBH_4$. *ent*-12β-Hydroxykaurenoic acid was incubated with a suspension culture of *Gibberella fujikuroi*, mutant B1-41a, at pH 4.8 for 5 days at 25°C (Gaskin *et al.*, 1984). Continuous arrows denote chemical synthesis; discontinuous arrows denote fungal conversions.

(1) Gas chromatography–mass spectrometry

Quantitative analysis by GC–MS is normally carried out using MIM. The mass spectrometer is set to jump to specific masses rather than to scan continuously over the entire mass range. Since the time allowed for collecting ions at each *m/z* value (the dwell time) is very much longer than during a full scan, MIM is a much more sensitive technique. Obviously the sensitivity increases when fewer ions are monitored, although the amount of information obtained is reduced. The advantage of MIM for quantitative analysis is its high sensitivity and selectivity. Although the same type of data could be obtained from mass chromatograms retrieved from the data system after full scanning, the sensitivity in this case would be about 10^2–10^3 times less.

MIM has been used by several workers to determine relative changes in GA concentration in comparable samples with time (Frydman *et al.*, 1974; Martin *et al.*, 1977; Ingram and Browning, 1979; Jones and Zeevaart, 1980; Metzger and Zeevaart, 1980; Yamaguchi *et al.*, 1982). In most cases several ions were monitored in order to confirm the identity of the GA that was being measured and to check that a particular peak was not distorted by ions from other components. Relative concentrations were based on the molecular ion except when this ion was too small to allow accurate determination of size (area or height). In this case a fragment ion was used. The technique has been called semiquantitative analysis (Yamaguchi *et al.*, 1982) since no account is taken of losses during extraction, purification and chromatography. Most workers expressed their results in relative terms, although Frydman *et al.* (1974) converted peak sizes to amounts from a calibration curve. This external standardization method is generally not suitable when extensive purification, which can result in substantial losses, is necessary.

Variation in the sensitivity of GC–MS can introduce significant errors unless it is taken into account. Ingram and Browning (1979) corrected for instrument variation by co-injecting with each sample a constant volume of a solution containing the GAs of interest derivatized with deuterated silylating reagent. Since they were measuring GA concentrations in *P. sativum* seeds they used an extract of this tissue as the source of the standard solution. Ions due to deuterated and non-deuterated derivatives were monitored simultaneously and the results were expressed relative to the deuterated derivatives. $[^{14}C]GA_3$ was used to calculate losses incurred during sample purification prior to GC–MS. This method allowed Ingram and Browning to estimate relative changes in GA concentration, although it was not possible to determine absolute values. It must be recognized that purification losses of all GAs are not accurately determined from the recovery of [^{14}C]GA_3. The method is also dependent on a careful injection technique for high precision.

(a) Use of an internal standard. Sample losses can be corrected for by including an internal standard in the extract immediately after the tissue is homogenized. The standard should be chemically as similar to the analyte as possible. For GC–MS, isotopically-labelled analogues of the analyte itself are used and original concentrations are estimated by isotope dilutions. Equivalent ions for the analyte and internal standard are monitored and the peak ratio for these ions is related to the weight ratio of the two species. GAs for use as internal standards can be labelled with ^{2}H, ^{13}C or ^{18}O. The production of labelled standards will be covered in Section VII.

Isotopically-labelled analogues are heavier than the analyte by an amount related to the number of isotopes that have been incorporated into the molecule. The difference in molecular weight between the internal standard and the analyte should be sufficiently large that there is no cross-contamination. Such contamination can occur due to the presence in the analyte of natural isotopes such as ^{13}C, ^{29}Si and ^{30}Si, which give rise to substantial M + 1 or M + 2 ions. On the other hand, if the labelled standard is not isotopically pure it may contribute to the size of the ion from the unlabelled species. The more atoms that are substituted, the less likelihood there is of unlabelled molecules being present.

The use of an internal standard will now be illustrated by an example. In order to determine the concentration of GA_4 in immature apple seeds, [$^{2}H_2$]GA_4 was prepared by the Wittig reaction (see Section VII). Although the reaction results in substitution at C-17, some scrambling of hydrogen atoms occurs during the reaction so that labelling at C-15 might also occur (Bearder *et al.*, 1976a). The mass spectra of the deuterated and non-deuterated GAs as MeTMS derivatives are shown in Fig. 17. A calibration curve can be constructed by mixing GA_4 and [$^{2}H_2$]GA_4 in different proportions and measuring peak areas from the MIM traces. Since GA_4 MeTMS has a very weak molecular ion, the intense fragment ions at *m/z* 284 and 286 were monitored for GA_4 and [$^{2}H_2$]GA_4 respectively.

The calibration curve is shown in Fig. 18*A*. The ratios of the peak areas for *m/z* 284 and 286 at the retention time of GA_4 MeTMS are plotted against molar ratios for the two species. A straight line can be fitted by the method of least squares. It can be seen that the line does not pass through the origin, the slope is not 1, actually ≈1.2, and the line deviates from linearity at high molar ratios. In Fig. 18*A* a straight line is fitted through only the first six points giving an acceptable correlation coefficient $r = 0.997$.

Assuming that there is not a significant isotope effect on the fragmentation to give the ion at *m/z* 286, the slope of the calibration line

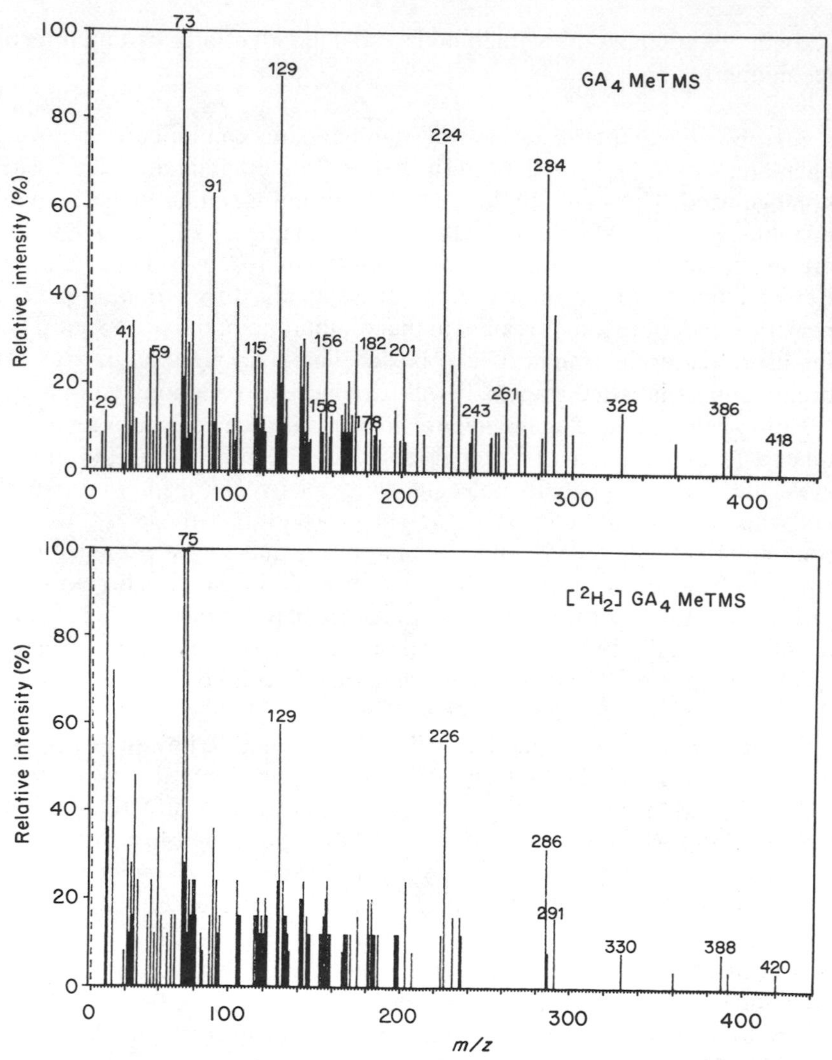

Fig. 17. Mass specta of GA_4 and $[^2H_2]GA_4$ as MeTMS derivatives. Mass spectrometer conditions as for Fig. 10.

indicates that the standard is not completely substituted. The intercept on the *y*-axis gives the contribution of the standard to the ion at *m/z* 284. This can be accounted for by the weak ion at *m/z* 282 in the mass spectrum of GA_4 MeTMS. Higher values would indicate the presence of completely unsubstituted GA_4 in the standard. The deviation of the line from linearity at high molar ratios shows a substantial contribution from the unlabelled GA_4 MeTMS to the ion at *m/z* 286. This contribution from natural isotopes can be calculated or measured and subtracted, or a higher-order regression curve could be fitted. The contribution from the unlabelled GA_4 MeTMS to the ion at *m/z* 286 was measured as 5% of the intensity of the *m/z* 284 ion. If this contribution is subtracted, a corrected calibration curve (Fig. 18*B*) is

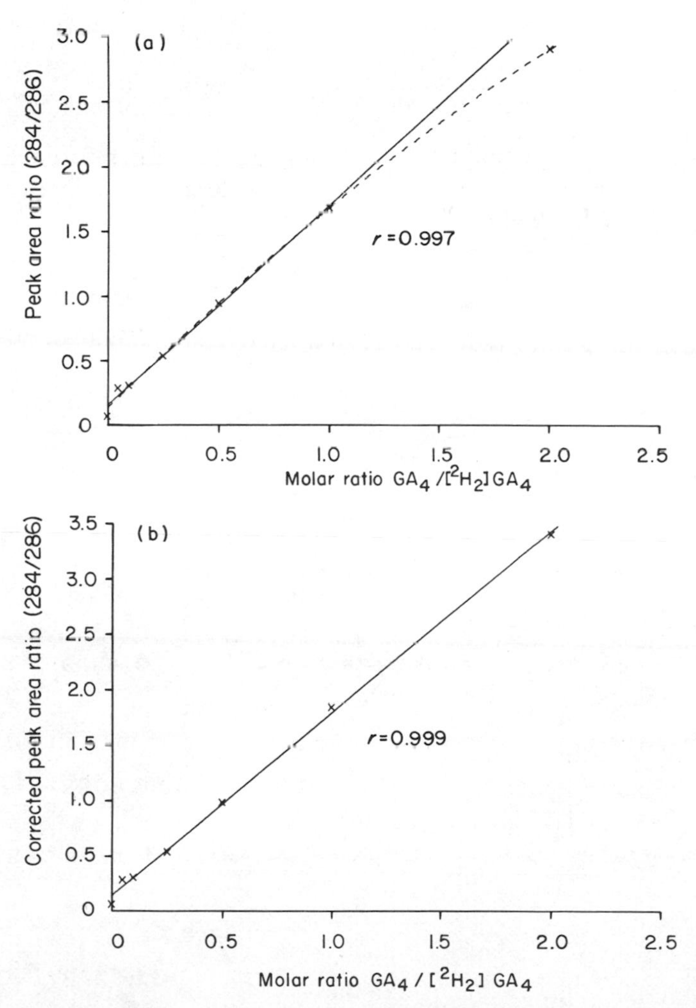

Fig. 18. Calibration curves for isotope dilution analysis of GA_4 by GC–MS using $[^2H_2]GA_4$ as an internal standard. Peak areas for the ions at *m/z* 284 and 286 were determined by GC-multiple ion monitoring. GC–MS conditions were as described for Figs 10 and 11. The dwell time for each ion was 50 ms. (a). Uncorrected calibration curve; the correlation coefficient is calculated from the first six points. (b). Corrected calibration curve after substracting 5% of the intensity of the *m/z* 284 ion from that of the *m/z* 286 ion.

obtained. In this case the correlation coefficient $r = 0.999$ for all seven points. Alternatively the upper portion of the curve can be made linear if an inverse plot, i.e. $[^2H_2]GA_4/GA_4$ against 286/284, is made.

An homogenate of 3.05 g of immature apple embryos was spiked with 3 μg $[^2H_2]GA_4$. After purification the derivatized extract was analysed by GC–MS. The MIM traces (Fig. 19) gave peak areas for *m/z* 284 and 286 of 175 and 103 units respectively. Thus the corrected peak area ratio

$$\frac{284}{286} = \frac{175}{(103 - 5.175/100)} = 1.86.$$

This corresponds to a molar ratio of 1.04 (Fig. 18*B*). Therefore the original extract contained 3.12 μg GA_4 corresponding to a tissue concentration of 1.02 μg g^{-1} fresh weight.

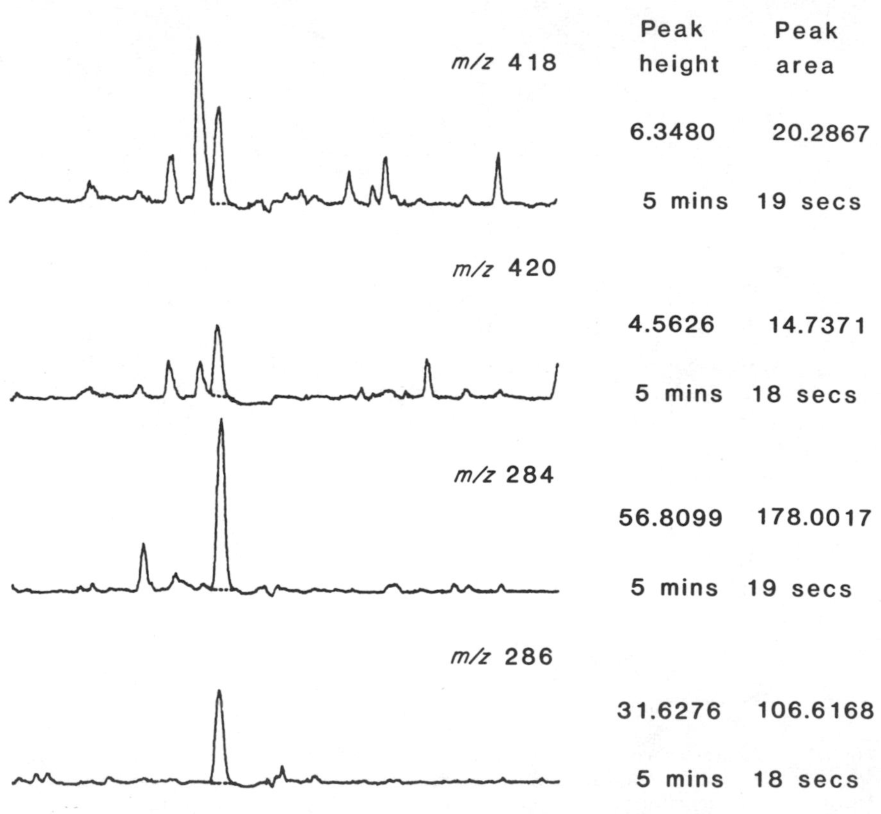

Fig. 19. GC-MIM traces for partially purified and derivatized (MeTMS) extract from embryos of *Malus*. The extract was spiked with $[^2H_2]GA_4$(1μg). GC–MS conditions were as described for Figs 10 and 11.

As an alternative approach, isotope dilution has been determined from complete mass spectra by measurement of ion intensities (Sponsel and MacMillan, 1977; 1978; 1980). This method has the advantage that the analyte is always adequately identified, but reasonably intense spectra are required for high precision. An additional problem when deuterated internal standards are used is that labelled and unlabelled species may separate slightly on the GC column. This does not matter when ion ratios are determined from MIM or mass chromatogram traces, but necessitates spectral averaging when ion intensities are calculated from the mass spectra (Sponsel and MacMillan, 1978).

Ideally, a labelled analogue of the GA to be measured should be used as internal standard. However, very few labelled GAs are available and most are extremely difficult to produce in sufficient quantities. This limitation can be partially overcome by using a single standard to measure several GAs. It should be recognized that large errors can result unless care is taken with the analysis. The standard should be chemically similar to the analyte, i.e. where possible a monohydroxylated C_{19}-GA should be measured relative to a labelled monohydroxylated C_{19}-GA. The purification procedure should not result in separation of the standard and analyte. The ions that are monitored for the standard and analyte are ideally as close in *m/z* value as possible. In order to express the results as absolute concentrations, the analyte must be available for the production of a calibration curve.

(b) Sources of error

(i) Producing the calibration curve can be the largest source of error. The variance at each point can be found by replicate injections of each solution and by the injection of replicate solutions. It should then be possible to determine whether the greater variance is due to GC–MS response or to making up the solutions. If the same standard solution is used for spiking samples as for the production of the standard curve any error in making up this solution can be disregarded.

(ii) Errors in weighing the original sample of plant material and measuring the volume of internal standard solution when spiking will contribute directly to errors in the final determination.

(iii) Labelled internal standards should be stable under the conditions used for extraction and purification. Deuterium atoms should be introduced at positions that are not exchangeable within the pH range (2.5–8.5) likely to be encountered. Labelling with ^{13}C would ensure stability. However, it is difficult to introduce more than a single ^{13}C atom, which would increase the molecular weight of the labelled GA by only 1 amu. This difference is not sufficient because of the large natural abundance at $M + 1$.

(iv) The MIM measurement can be seriously affected by interfering ions from other components in the extract. It is useful to monitor several ions for both analyte and standard in order to discount this possibility and to confirm the identities of species being measured.

(v) The ratio of analyte to standard should lie on a linear part of the calibration curve. This may involve two determinations; the first to obtain an estimate of the GA concentration, which is then determined more accurately in the second.

(vi) The efficiency of extraction of the hormone from the plant tissue is difficult to ascertain. Successive extractions should ensure that all extractable GAs under the conditions employed have been removed.

The estimation of error in isotope dilution GC–MS is beyond the scope of this chapter. The reader is referred to a paper by Moler *et al.* (1983) for a full discussion of this subject.

(2) Immunoassays

The introduction of immunoassays to the analysis of plant growth substances has been one of the most important innovations in this field in recent years. However, due to the large qualitative differences in GA molecular species that are encountered in plants, the application of immunoassays to GA analysis is more complex than for other classes of growth hormone. Fuchs and co-workers were the first to apply immunological techniques to GA analysis, but the methods were insensitive and unselective (Fuchs and Fuchs, 1969; Fuchs *et al.*, 1971; Fuchs and Gertmann, 1974). Weiler and co-workers have now developed extremely sensitive techniques that compare favourably with GC–MIM (Weiler and Wieczorek, 1981; Atzorn and Weiler, 1983a; 1983b).

The principle of the immunoassays described by Weiler *et al.* is to allow the analyte to compete with a tracer for antibody binding sites. The tracer should thus have a similar affinity for the binding sites as the analyte. After equilibrium has been attained, bound and unbound species are separated and the amount of bound tracer is measured. This amount is inversely related to the amount of analyte present, which can then be determined by reference to a calibration curve. A typical calibration curve is shown in Fig. 20, reproduced from Weiler and Wieczorek (1981). It is produced by incubating constant aliquots of antibody and tracer with a range of concentrations of a standard. Then the percentage of the tracer bound in the presence of the standard (B) to that bound in its absence (B_0) is plotted against the log of the standard concentration. The curve is sigmoidal, but the centre portion can be linearized using the logit transformation (Fig. 20):

$$\text{logit } B/B_0 = \ln [(B/B_0)/(100 - B/B_0)].$$

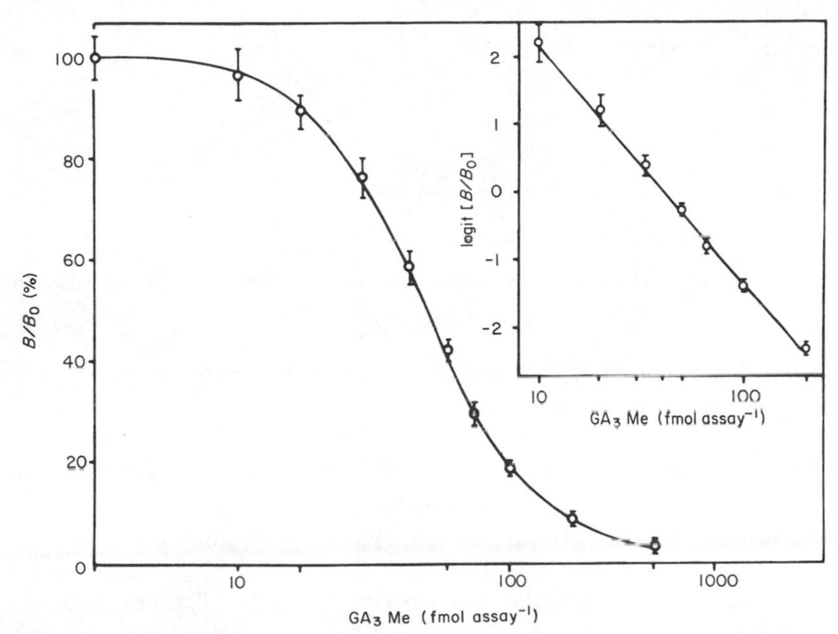

Fig. 20. Calibration curve for the radioimmunoassay of GA_3 methyl ester using the derivative (**8**) as tracer. B, tracer bound in the presence of GA_3; B_0, tracer bound in its absence. Inset, linearized calibration curve produced using the logit transformation as described in the text. Reproduced, with permission, from Weiler and Wieczorek (1981).

The precision of quantitative estimates depends on the steepness of the calibration curve.

GAs, in common with the other classes of plant growth substances, are too low in molecular weight to be highly immunogenic: they have to be coupled to macromolecules in order for antibodies to be raised. Antibodies against GA_3 (Weiler and Wieczorek, 1981; Atzorn and Weiler, 1983a; 1983b), GA_1 (Atzorn and Weiler, 1983a), GA_9 (Atzorn and Weiler, 1983a) and $GA_{4/7}$ (Atzorn and Weiler, 1983b) have been raised in rabbits using as immunogens the GAs coupled through the C-7 carboxyl group to bovine serum albumin (BSA). Two coupling procedures were employed. In one, a mixed anhydride between the GA and isobutylchloroformate was formed and then allowed to react with BSA (Weiler and Wieczorek, 1981) (Fig. 21). Alternatively the GA anhydride, formed by reaction with dicyclohexylcarbodiimide, was reacted with BSA or *p*-aminohippuric acid-substituted BSA (Atzorn and Weiler, 1983b). The degree of coupling was estimated as 4–8 molecules of GA per molecule of protein, and high-affinity antibodies were raised. The K_a of each antiserum for binding of the respective antigen as the Me ester was determined from a Scatchard analysis as over 10^{10} l mol^{-1}. The

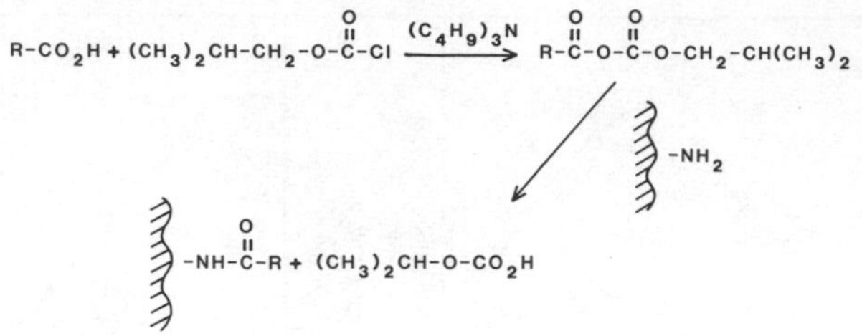

$$R\text{-}CO_2H + (CH_3)_2CH\text{-}CH_2\text{-}O\text{-}\overset{O}{\overset{\|}{C}}\text{-}Cl \xrightarrow{(C_4H_9)_3N} R\text{-}\overset{O}{\overset{\|}{C}}\text{-}O\text{-}\overset{O}{\overset{\|}{C}}\text{-}O\text{-}CH_2\text{-}CH(CH_3)_2$$

$$\text{-}NH_2$$

$$\text{-}NH\text{-}\overset{O}{\overset{\|}{C}}\text{-}R + (CH_3)_2CH\text{-}O\text{-}CO_2H$$

Fig. 21. Procedure for coupling a GA to protein via a mixed anhydride with isobutylchloroformate.

antisera had much higher affinities for the GA Me esters than for the free acids.

(a) Radioimmunoassay. Radiolabelled tracers are easily detected and quantified. The sensitivity of the assay depends both on the affinity of the antiserum for the antigen and on the specific radioactivity of the tracer. Weiler and Wieczorek (1981) produced an ^{125}I-labelled derivative of GA_3 (**8**) of extremely high specific radioactivity (8.5×10^{16} Bq mol^{-1}), but this tracer was unstable and difficult to synthesize in good yield and had a radioactivity half-life of only two months. Atzorn and Weiler (1983a) changed to using tritium-labelled GAs as tracers. These were prepared from the 17-norketones (Section VII) by reduction with NaB^3H_4 and had specific radioactivities estimated at *ca.* 2×10^{14} Bq mol^{-1}.

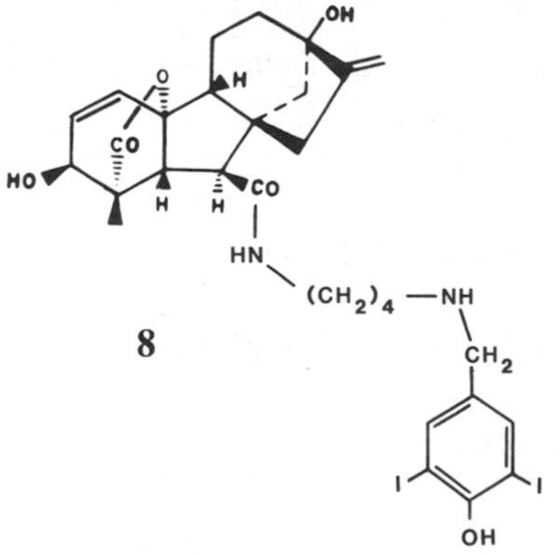

8

The radioimmunoassay (RIA) was carried out by incubating an aliquot of antiserum with the tracer and a portion of plant extract or a known amount of a standard. The aliquot of antiserum was chosen to bind 35–50% of the tracer in the absence of standard or extract. After incubating for several hours at 4°C, bound and unbound antigens were separated by precipitating the antibodies with ammonium sulphate and centrifuging. The formation of a pellet can be facilitated by including some non-antigenic protein in the incubation mixture. The supernatant was removed and the amount of radioactivity in the pellet was determined by liquid scintillation counting.

Since the antisera had higher affinities for the Me esters, the tracers, standards and extracts were methylated prior to analysis. Detection limits of 70 fmol for GA_1 and 100 fmol for GA_3 and GA_9 were obtained.

(b) Enzyme-linked immunoabsorbent assay. In an enzyme-linked immunoabsorbent assay (EIA) an enzyme–analyte conjugate is used as tracer. The amount of bound tracer is measured via a reaction catalysed by the enzyme, allowing for considerable amplification. EIAs can therefore be extremely sensitive. Atzorn and Weiler (1983b) used GA-alkaline phosphatase conjugates as tracers. GA_3 and GA_4 were first converted to the anhydrides in the presence of equimolar amounts of dicyclohexylcarbodiimide, and the anhydrides were allowed to react with the enzyme. The

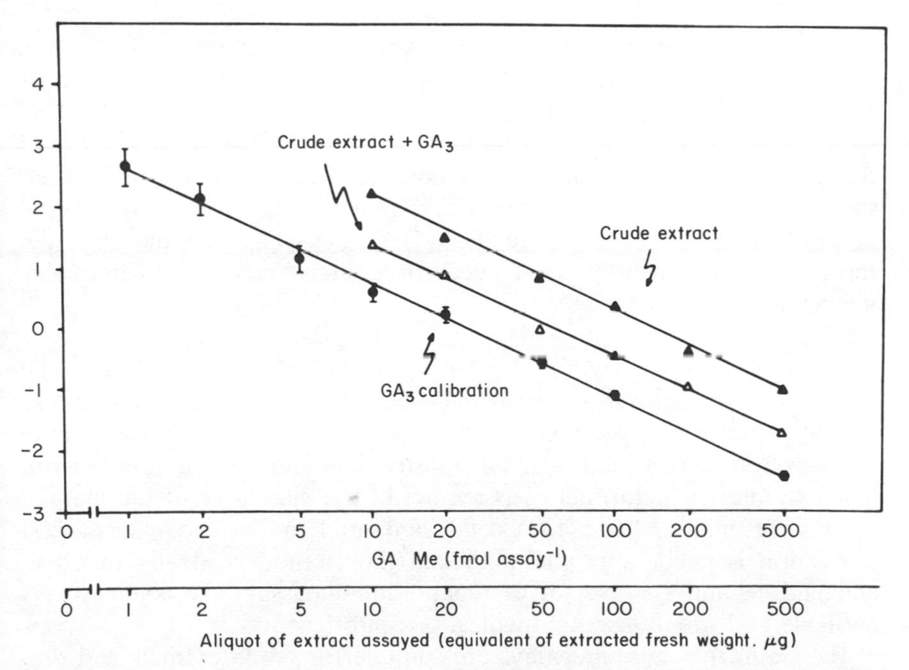

Fig. 22. Comparison of logit binding curves obtained using pure standard or using aliquots of an acidic ethyl acetate fraction from extracts of *Hordeum vulgare*. The curves were obtained using a solid-phase enzyme assay with GA_3-alkaline phosphatase as tracer. Reproduced, with permission, from Atzorn and Weiler (1983b).

immunoassay was carried out on polystyrene microtest plates, which were first coated with the antibody. The precision of the assay is limited by the density of antibody bound to the plate. In order to increase the coating density, Atzorn and Weiler used purified IgG fractions prepared from the antiserum by ammonium sulphate precipitation and ion-exchange chromatography. After the coated plates were incubated with tracer and standards or aliquots of extract, unbound material was removed and the amount of bound tracer was determined colorimetrically using *p*-nitrophenyl phosphate as substrate. The absorption at 405 nm due to the reaction product (*p*-nitrophenol) was measured ($\varepsilon_{405} = 18.5 \times 10^6 \, cm^2 mol^{-1}$). Calibration curves for GA Me esters can be produced using the logit transformation in the same way as in the radioimmunoassay. Although slightly less precise than the radioimmunoassay, the enzyme assay was about two orders of magnitude more sensitive. The detection limit for GA_3, using antiserum raised against GA_3-BSA and GA_3-alkaline phosphatase as tracer was 0.5 fmol. GA_4 and GA_7, assayed using antiserum against $GA_{4/7}$-*p*-aminohippuric acid-coupled BSA and GA_4-alkaline phosphates as tracer had detection limits of 1.0 and 1.5 fmol respectively.

(c) Specificity of antisera. The cross-reactivity of methylated GAs and related compounds with five antisera as determined by Atzorn and Weiler (1983a; 1983b) are listed in Table VIII. The data are expressed as the relative amounts of standard and competing compound that displace 50% of the tracer. It can be seen that the antisera differ considerably in their specificity. The antiserum against GA_9 is quite specific, whereas that raised against GA_1 binds a broad range of C_{19}-GAs. Thus it appears that structure recognition by the antibody depends to a large extent on hydrophobic interactions.

(d) Critical assessment of immunoassays for gibberellin analysis. The main advantages of immunoassay over most other analytical methods are high sensitivity and ease of execution. Unlike GC–MS it need not require expensive equipment although some instrumentation, e.g. a liquid scintillation counter or colorimeter, is required. The simplicity of the method enables very many analyses to be conducted simultaneously so that sufficient replication is never a problem. Indeed the method is ideally suited to automation, and systems for automatic immunoassays are commercially available and widely used in medical laboratories.

The claim that immunoassays are suitable for crude extracts and thus obviate the need for extensive sample purification requires careful examination. It is important to establish that components of the extract do not interfere with antigen–antibody binding. In order to discount interference,

Table VIII. Cross-reactivities of GAs and related compounds with antisera raised against GAs. The results are expressed on a % molar basis. All compounds were methylated before analysis.

	Antiserum raised against				
Compound tested	GA_1[a]	GA_3[a]	GA_3[b]	$GA_4 + GA_7$[b,c]	GA_9[a]
GA_1	100	11	13	13	<0.1
GA_3	70	100	100	<0.1	<0.1
GA_4	40	9	7	100	3
GA_5	29	0.2	<0.1	0.1	<0.1
GA_7	70	35	40	80	1.1
GA_8	11	<0.1	<0.1	0.5	<0.1
GA_9	15	<0.1	<0.1	<0.1	100
GA_{12}	<0.1	<0.1	<0.1	<0.1	<0.1
GA_{13}	<0.1	<0.1	<0.1	<0.1	<0.1
GA_{16}	0.1	<0.1	<0.1	<0.1	<0.1
GA_{19}	<0.1	<0.1	0.1	<0.1	<0.1
GA_{20}	55	22	22	1.5	<0.1
GA_{24}	<0.1	<0.1	<0.1	<0.1	<0.1
GA_{32}	<0.1	<0.1	<0.1	3.5	<0.1
GA_{34}	<0.1	<0.1	<0.1	1.8	<0.1
GA_{42}	<0.1	<0.1	<0.1	<0.1	<0.1
iso-GA_3	10	<0.1	<0.1	0.2	<0.1
iso-GA_7	3	<0.1	<0.1	11	<0.1
steviol (**27**)	0	0	0	0	0
GA_8-2*β*-*O*-glucoside	0	0	0	0	0

[a]Determined by radioimmunoassay using [^{3}H]GA as tracer (Atzorn and Weiler, 1983a); [b]determined by enzyme immunoassay using GA-alkaline phosphatase as tracer (Atzorn and Weiler, 1983b); [c]GA_4-alkaline phosphatase used as tracer.

Atzorn and Weiler compared logit binding curves in the presence of different amounts of sample extract with the standard curve (Fig. 22). The curves should be parallel. It has, however, been found that parallel logit curves can sometimes be obtained despite considerable interference (H. G. Jones, unpublished results). A better method is to measure antigen concentrations in several dilutions of the extract in the presence of different amounts of exogenous antigen (Rosher *et al.*, 1985). The results should be a series of straight lines, which, in the absence of interference, will be parallel. Atzorn and Weiler also compared results from crude extracts with those obtained after purification by TLC. In this case the extracts were spiked with [^{3}H]GA to correct for losses. Since interference will be a problem with many plant extracts, some purification, requiring the inclusion of internal standards, may be necessary.

Separation of the different GAs in an extract by, for example, HPLC will always be necessary unless antisera of very high specificity are available. The cross-reactivity data in Table VIII indicate that such antisera may be difficult to produce by the methods described above. Only the GA_9 antiserum has the necessary specificity. Unless the cross-reactivities of all possible GAs that may co-chromatograph with the analyte are known to be negligible, it must be shown by some other method that the extract or fraction contains a single immunoreactive component. This task should become easier as the cross-reactivities of more of the known GAs are determined with the available antisera. Furthermore, two or more antisera with different GA specificities may be used to check the validity of a determination. However, as a general rule it is preferable to establish the qualitative composition of a system before quantitative analyses are attempted.

(e) Monoclonal antibodies. High antibody specificity is often difficult to achieve since in response to immunization with an antigen, many antibodies of different affinities and specificities are produced. Homogeneous antibodies of potentially high affinity and specificity can be obtained using monoclonal antibody technology (Melchers *et al.*, 1978). Simply stated, spleen lymphocytes from immunized animals are fused with cells from a myeloma cell line. The resulting hybrids are cloned and the clones are screened for the production of antibodies of the required specificity. The cloned hybrid cells can be stored and grown in culture when antibody is required. Since the cells are selected it is not necessary to immunize with pure antigen. In fact, several cell lines producing different antibodies after immunization with a mixture of antigens could be raised at the same time.

Monoclonal antibodies (McAB) to GAs have now been raised in at least two laboratories. Eberle *et al.* (1986) obtained two antibodies derived from mice immunized with a bovine serum albumin (BSA) conjugate of GA_{13}-19,20-imide-β-alanine 7-methyl ester (**9**). These McAB exhibited high affinities for GA_4 methyl ester and allowed quantitation of this GA, and several others, by radioimmunoassay in the subnanogram range after methylation. Knox *et al.* (1987) used the keyhole limpit haemocyanin (KLH) conjugates **10–12** to produce six rat-derived McAB with high affinities for several GAs as free acids. The detection ranges were also in the subnanogram region. Some characteristics of the McAB are given in Table IX. None of the antibodies is monospecific, but by conjugating GAs at different positions in the molecule Knox *et al.* were able to produce antibodies specific to particular epitopes. For example, MAC 136, which was produced using a conjugate in which the GA-hapten is coupled via C-3, requires a hydroxyl group on C-13 for binding, but is relatively unspecific for substitution on the A ring. MAC 183, which was produced using GA_4 coupled via the exocyclic methylene group to the protein, binds well only if the C-3 hydroxyl group is present and is relatively tolerant of a C-13 hydroxyl.

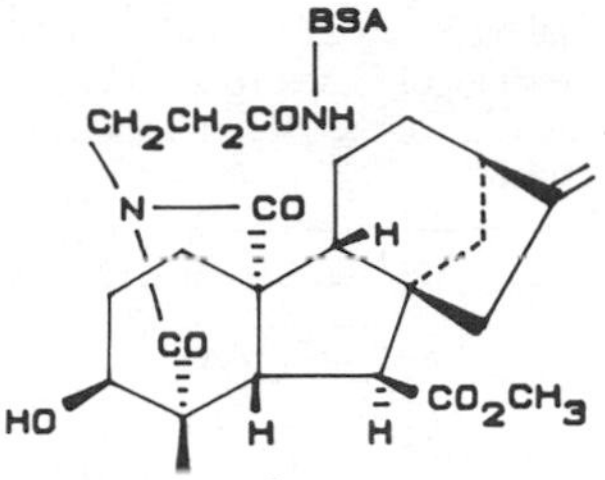

9

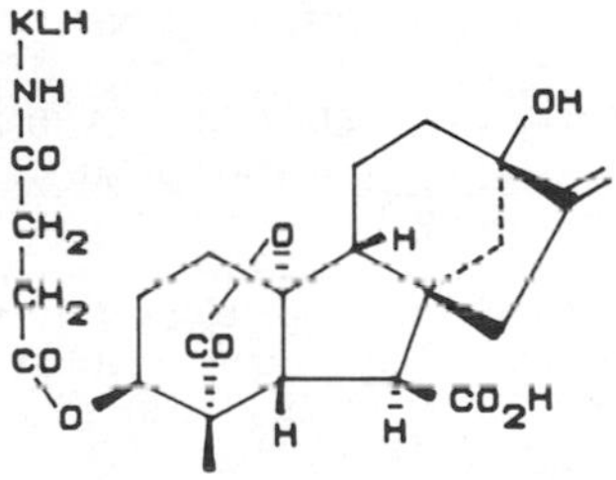

10

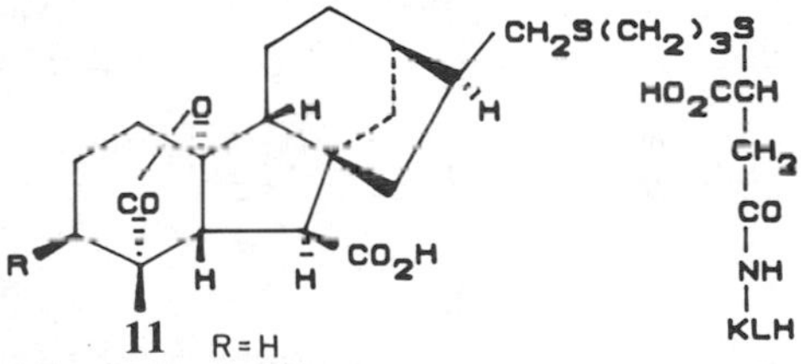

11 R=H
12 R-OH

It is clear, at least with the antibodies available so far, it is not possible to identify, and therefore quantify, a particular GA with a single McAB in a crude plant extract. However, the combined use of several McAB would provide considerable structural information about an unknown GA. The advantage of McAB over polyclonal antibodies is that the hybridoma cell lines provide an unlimited supply of antibody of defined properties. Perhaps one of the most important applications of GA antibodies is in immunoaffinity chromatography, which could provide a rapid one-step purification. Antibodies also have potential for use in GA immunolocalization in plant tissue at the cellular and sub-cellular level. This technique has been explored already for some of the other plant hormones (see for example Zavala and Brandon, 1983).

V. ANALYSIS OF GIBBERELLIN CONJUGATES

GAs are found covalently linked to a limited number of low molecular-weight molecules. These so-called GA conjugates are of widespread occurrence and accumulate to high concentrations in mature seeds of certain species (Schneider, 1983). By far the most abundant of these compounds are GA–glucose conjugates. Other derivatives, such as low molecular-weight esters (Hemphill *et al.*, 1973) and gibberethione, a complex adduct of GA_3 with an amino acid (Yokota *et al.*, 1974), have also been reported.

Table IX. Properties of monoclonal antibodies (McAb) to GAs. To provide an indication of specificity the cross-reactivities of a range of GAs are given. These were determined from the molar amounts required to inhibit binding of the [^{3}H]GA used as tracer by 50% and expressed relative to the tracer GA on a % basis.

McAb (class)	Immunogen	Affinity M^{-1}	Cross Reactivities
J51-II-C1[a] (IgG1)	**9**	1.0×10^9	GA_1(1)[b]; GA_2(46); GA_3(37); GA_4(100); GA_7(805); GA_9(26); GA_{20}(2); GA_{34}(15); GA_{13}(7); GA_{14}(<0.3); GA_{37}(360); GA_{53}(3)
J51-III-A4[a] (IgG1)	**9**	3.2×10^8	Ga_1(1)[b]; GA_2(42); GA_3(76); GA_4(100); GA_7(764); GA_9(53); GA_{20}(2); GA_{53}(4)
AFRC MAC 136[c] (IgM)	**10**	7.5×10^7	GA_1(100)[d]; GA_2(0.07); GA_3(100); GA_4(0.4); GA_5(48); GA_7(0.3); GA_8(53); GA_9(0.2); GA_{20}(100); GA_{29}(100); GA_{14}(0.7); GA_{53}(167)
AFRC MAC 137[c] (IgG2c)	**10**	9.2×10^7	GA_1(100)[d]; GA_2(0.08); GA_3(100); GA_4(0.5); GA_5(67); GA_7(0.4); GA_8(80); GA_9(0.5); GA_{20}(100); GA_{29}(100); GA_{53}(180)
AFRC MAC 175[c] (IgM)	**11**	1.2×10^8	GA_1(<0.02)[e]; GA_2(0.08); GA_3(<0.02); GA_4(0.1); GA_5(3); GA_7(0.07); GA_8(<0.02); GA_9(100); GA_{20}(16); GA_{29}(<0.5); GA_{14}(<0.04); GA_{53}(0.5)
AFRC MAC 176[c] (IgM)	**11**	4.7×10^9	GA_1(6)[e]; GA_2(30); GA_3(1); Ga_4(100); GA_5(2); GA_7(11); GA_8(0.06); GA_9(100); GA_{20}(9); GA_{29}(<0.2); GA_{14}(0.5)
AFRC MAC 182[c] (IgG2a)	**12**	1.1×10^{10}	GA_1(48)[f]; GA_2(50); GA_3(6); GA_4(100); GA_5(0.8); GA_7(9); GA_8(0.03); GA_9(0.9); GA_{20}(0.5); GA_{29}(0.02); GA_{14}(0.01); GA_{53}(0.03)
AFRC MAC 183[c] (IgM)	**12**	8.7×10^7	GA_1(50)[f]; GA_2(24); GA_3(4); GA_4(100); GA_5(2); GA_7(15); GA_8(0.2); GA_9(4); GA_{20}(3); GA_{29}(<0.06); GA_{14}(<0.02)

[a]Eberle *et al.* (1986); [b][1,2-$^3H_2$]GA_4 Me ester was used as tracer for determining affinity and cross-reactivities, all GAs were methylated prior to cross-reactivity determination; [c]Knox *et al.* (1987) [d][1,2-$^3H_2$]GA_1 was used as tracer for determining affinity and cross-reactivities; [e][2,3-$^3H_2$]GA_9 was used as tracer for determining affinity and cross-reactivity; [f][1,2-$^3H_2$]GA_4 was used as tracer for determining affinity and cross-reactivities.

A. Purification

The GA–glucose conjugates are found as *O*-*β*-glucopyranosyl ethers and esters. The glucose ethers, coupled through a GA hydroxyl group, are highly water-soluble acids, whereas the esters, which are coupled to the GA 6*β*-carboxyl group, are neutral, but polar, molecules. In the purification scheme in Fig. 6, GA-glucosyl ethers are found in the butanol fraction after partition at pH 2.5. The glucosyl esters, however, may be distributed between the butanol and neutral ethyl acetate fractions. Hiraga *et al*. (1974) in their isolation of GA–glucosyl esters from mature *Phaseolus vulgaris* seeds, found that the glucosyl esters of the monohydroxy GAs, GA_4 and GA_{37}, were located exclusively in the neutral (pH 7.0) ethyl acetate extract, whereas GA_1- and GA_{38}-glucosyl esters were in both neutral ethyl acetate (10%) and acidic butanol (90%) fractions. Schneider (1983) has pointed out that, because of the influence of the polar glucose moiety, the partition characteristics of GA-glucosyl derivatives are not as pH-dependent nor as predictable as they are for the free GAs. He suggested that in order to avoid dispersion during partition, a total conjugate fraction be extracted with butanol at low pH and subsequent purification be afforded by ion-exchange chromatography, for example on DEAE-Sephadex (Gräbner *et al.*, 1976). Yamaguchi *et al.* (1980) used the scheme outlined in Fig. 23 to isolate GA_5- and GA_{44}-glucosyl esters from immature *Pharbitis purpurea* seeds. The partition sequence would be expected to confine the esters to the neutral butanol fraction.

In general the methods used for GA-conjugate purification are similar to those applied to the free acids. Thus columns of charcoal and silica gel as well as Sephadex G-50 partition chromatography were used by Hiraga *et al.* (1974). However, such polar molecules as GA conjugates will be recovered poorly from charcoal or silica gel adsorption columns. Reversed-phase HPLC has been applied to GA-conjugates and promises to be an extremely useful technique (Yamaguchi *et al.*, 1979; 1980; Koshioka *et al.*, 1983a). Yamaguchi *et al.* (1979) found that reasonable separation of several GA-conjugates could be achieved on Wakogel LC ODS eluted isocratically with 15% methanol in 10 mM ammonium chloride at pH 3.2. They used this system to purify GA_3-3-glucoside from *Quamoclit pennata* seeds. Nucleosil C_{18} eluted isocratically with 30% 2-propanol/water or 50% methanol/water was used by Yamaguchi *et al.* (1980) to purify GA_5- and GA_{44}-glucosyl esters. Table X lists the retention times for several GA-conjugates obtained by Jensen *et al.* (1986) on a Supelcosil analytical (250 × 4.6 mm i.d.) column eluted isocratically with methanol in aqueous phosphoric acid.

B. Identification

Although there have been numerous reports of the detection of GA-conjugates in plant tissues, in few cases has the conjugate itself been isolated and characterized. In most cases the glucoside has been hydrolysed and the free GA identified or quantified. This method simplifies the analysis, but provides little information on the position of conjugation. Furthermore, since the appropriate labelled GA conjugate is seldom available as an internal standard it is not possible to estimate the efficiency of hydrolysis so that quantitation is unreliable.

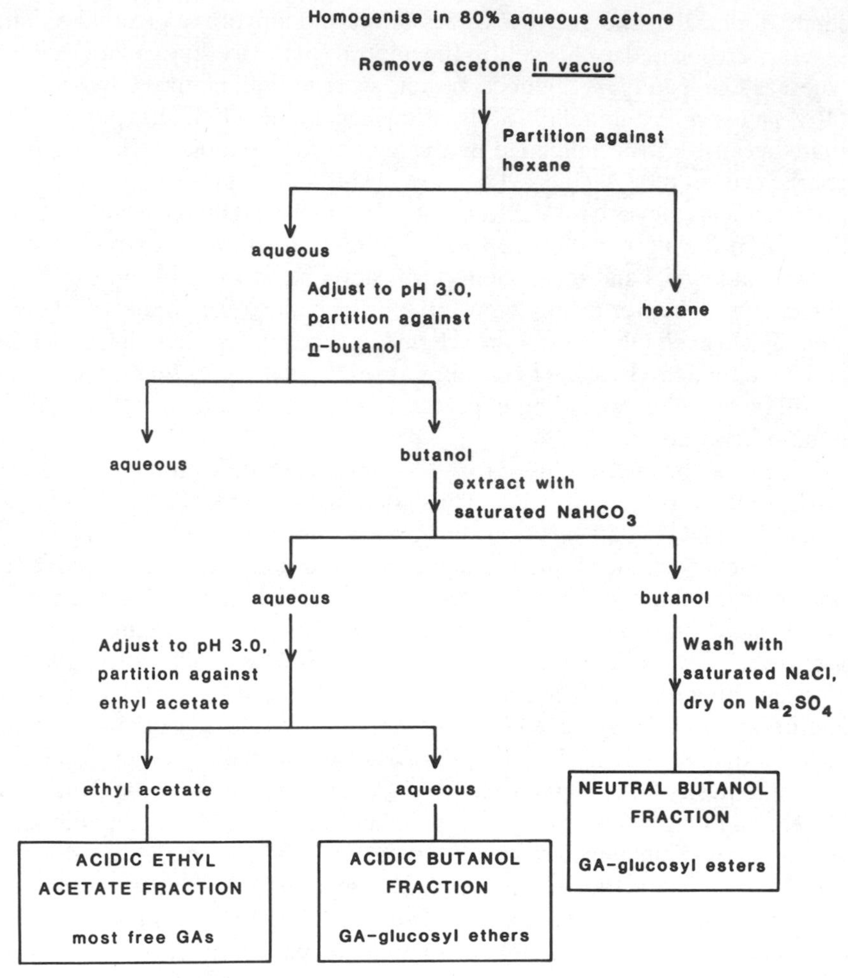

Fig. 23. Solvent–solvent partition scheme used by Yamaguchi *et al.* (1980) to purify GA-glucosyl esters.

Since most GAs are unstable at extremes of pH, it is usually necessary to cleave the glucoside by enzymatic hydrolysis. Several enzymes have been used for this purpose. In a comparative study, Müller *et al.* (1978) found cellulase (*Aspergillus niger*) and helicase (*Helix pomatia*) to be the most effective. Cellulase, pectinase and β-glucosidase are used most commonly. They are incubated with the extract in buffered solution at pH 4.5 for 12–24 h at about 37°C. The rate of hydrolysis depends on the position of attachment of the glucose moiety. GA-2β-*O*-glucosides, in which glucose is attached to an equatorial hydroxyl group, are rapidly hydrolysed, as are

Table X. Reversed-phase HPLC retention properties of gibberellin glucosyl conjugates (Jensen *et al.*, 1986). Gibberellin conjugates analysed on a 250 × 4.6 mm i.d. 5 μm Supelcosil LC 18 column eluted isocratically at 1 ml min^{-1} with methanol in aqueous phosphoric acid at pH 3.0. Detection with an UV absorbance monitor at 208 nm. Data expressed as retention times (min).

	Methanol (%)									
	10	15	20	25	30	35	40	45	50	55
GA_{29}-2β-*O*-glucoside	19.9	11.2	7.4							
GA_8-2β-*O*-glucoside	23.8	12.6	8.0							
Gibberellenic acid-2β-*O*-glucoside		25.7	13.6	8.9						
GA_3-13-*O*-glucoside		26.1	13.9	8.9						
GA_1-13-*O*-glucoside		30.0	15.9	10.0						
GA_3-3β-*O*-glucoside			22.1	12.8	8.4					
GA_1-3β-*O*-glucoside			23.3	13.5	8.7					
GA_{38}-glucosyl ester			23.9	14.5	9.2					
GA_{26}-2β-*O*-glucoside			25.1	15.6	9.6					
GA_3-glucosyl ester			26.4	16.0	9.8					
GA_1-glucosyl ester			31.1	17.8	11.2					
GA_{35}-11β-*O*-glucoside				19.6	11.3	8.0				
GA_5-13-*O*-glucoside					24.4	14.5	9.2			
GA_{20}-13-*O*-glucoside					25.2	15.2	9.8			
GA_5-glucosyl ester					44.9	23.1	13.1			
GA_{20}-glucosyl ester						24.3	14.0	9.4		
GA_{37}-glucosyl ester						25.3	17.3	11.2		
GA_7-3β-*O*-glucoside							20.1	12.6	8.4	
GA_7-glucosyl ester							24.5	15.3	9.8	
GA_4-glucosyl ester								22.6	13.6	8.9

13-*O*-glucosides, whereas the axial 3β-*O*-glucosides such as in GA_1- and GA_4-glucosides are hydrolysed relatively poorly (Schneider and Schliemann, 1979). GA_3-3β-*O*-glucoside and GA_7-glucoside are hydrolysed by cellulase about four times more rapidly than GA_1- and GA_4-glucosides. Thus the effectiveness of enzyme treatment will depend on the nature of the GA-glucosides.

Chemical characterization of GA-glucosides has been undertaken when sufficient compound was isolated (Yokota *et al.*, 1971; Schreiber *et al.*, 1969), but this has involved the extraction of enormous quantities of plant material. Schreiber *et al.* isolated 800 mg of GA_8-2-*O*-glucoside from 3762 kg of *Phaseolus coccineus* pods. The use of sensitive analytical techniques such as GC–MS has been limited by the involatility of the glucosides. TMS derivatives can be gas chromatographed, although they require high temperatures at which decomposition can occur, and have long retention times (Schneider *et al.*, 1975). Abbreviated mass spectra of some TMS derivatives are presented in Table XI, which lists the 10 most significant ions and the molecular ion. The data are taken from Yokota *et al.* (1975). The TMS ethers and Me esters give very large molecular ions, which in many instances are extremely weak. Cleavage tends to occur preferentially on the glucose moiety. The glucose esters produce intense ions at M^+-378, corresponding to the GA TMS ester. Although the combination of mass spectrum and GC retention time may be used to identify known GA–glucosides, the mass spectra of TMS derivatives give little structural information that might help in the identification of an unknown conjugate, apart from the molecular weight of the aglycone.

The possibility of using permethylated GA-conjugates for GC–MS was examined by Rivier *et al.* (1981). The method was discussed in Section IV. The permethylated conjugates have much lower molecular weights than the TMS derivatives and can be gas chromatographed without decomposition. The EI mass spectra give extremely weak or absent molecular ions and are dominated by fragments from the glucose moiety (see for example Fig. 24). Schneider (1983) has pointed out that negative ion EI–MS produces relatively intense molecular ions for GA-glucoside Me esters and peracetylated Me esters, the latter being potentially useful for GC–MS analysis.

Table XI. Molecular ion and 10 most intense fragment ions in the mass spectra of GA-glucosides MeTMS[a] and GA-glucosyl esters TMS[b].

Compound	M^+	*m/z* (% relative intensity)				
GA_3-3β-*O*-glucoside	882(28)	810(25)	443(34)	431(100)	415(36)	387(24)
		371(50)	370(40)	369(28)	311(38)	281(28)
GA_8-2β-*O*-glucoside	972(4)	623(57)	551(91)	550(28)	533(35)	505(50)
		478(41)	461(100)	401(34)	373(30)	372(59)
GA_{26}-2β-*O*-glucoside	898(t)	549(29)	505(18)	490(18)	477(100)	459(19)
		431(26)	387(48)	327(23)	313(19)	297(32)
GA_{27}-2β-*O*-glucoside	898(1)	549(52)	477(100)	448(16)	431(27)	387(69)
		374(17)	359(54)	327(19)	299(27)	283(57)
GA_{29}-2β-*O*-glucoside	884(18)	535(93)	503(38)	489(55)	462(30)	417(100)
		385(35)	373(68)	357(29)	327(41)	313(30)
GA_{35}-11-*O*-glucoside	884(t)	535(100)	506(43)	489(19)	445(21)	417(74)
		399(35)	371(25)	361(90)	327(27)	283(65)
GA_1-glucosyl ester	942(11)	606(10)	564(100)	547(10)	492(50)	491(13)
		477(19)	476(28)	475(20)	447(33)	415(9)
GA_3-glucosyl ester	940(9)	604(25)	562(100)	490(90)	489(21)	475(27)
		474(27)	473(25)	445(60)	401(27)	355(25)
GA_4-glucosyl ester	854(3)	476(100)	461(22)	459(42)	431(24)	388(26)
		387(72)	386(45)	359(30)	341(26)	269(44)
GA_{37}-glucosyl ester	868(1)	724(8)	490(100)	475(7)	446(19)	418(10)
		402(9)	401(28)	373(8)	328(8)	283(17)
GA_{38}-glucosyl ester	956(2)	578(100)	563(14)	506(33)	491(34)	490(62)
		489(34)	461(32)	417(20)	416(18)	371(35)

Data reproduced from Yokota *et al.* (1975). [a]Only ions >M^+-615 included; [b]only ions >M^+-585 included; t, trace.

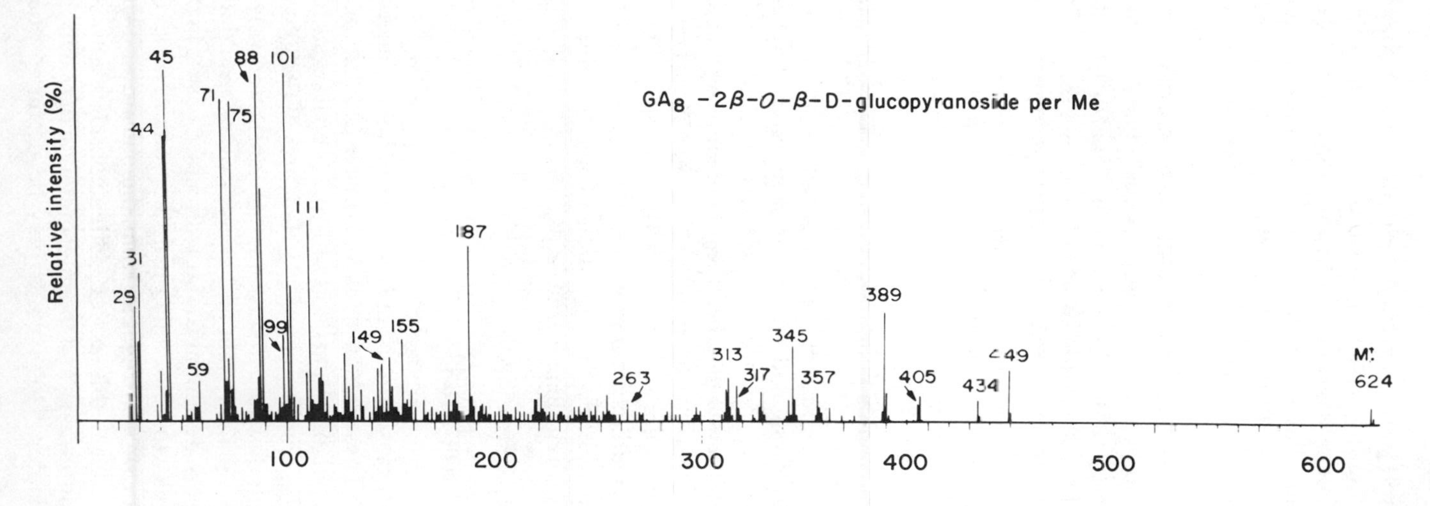

Fig. 24. Electron impact mass spectrum of the permethylated derivative of GA_8-2β-*O*- glucoside. The spectrum was obtained using a MS30 mass spectrometer at 24 eV. Reproduced, with permission, from Rivier *et al.* (1981).

Analytical HPLC will prove of great benefit for involatile GA derivatives such as glucosides. The full power of this technique will probably not be realized until the problems associated with coupling liquid chromatographs to mass spectrometers have been overcome. At present the detectors available with HPLC are not sufficiently selective for this method to be used directly as an analytical technique rather than as an extremely useful purification step.

Quantitative analysis of GA-conjugates has been restricted to GA determination in hydrolysed extracts. Direct physicochemical methods require internal standards, preferably with an isotopic label. Chemical methods have been developed for the synthesis of GA–glucosyl esters (Schneider *et al.*, 1977) and esters (Schneider *et al.*, 1984). The general method based on the Koenigs–Knorr reaction is outlined in Fig. 25. The most difficult aspect is removal of the acetate protecting groups without affecting the labile GA molecule. Due to steric effects, the yield of glucosidation at C-3 is low (Schneider, 1980).

Turnbull *et al.* (1986) recently described a tentative identification of radiolabelled GA-conjugates resulting from a feed of $[^3H]GA_4$ to *Phaseolus vulgaris* seedlings. GA conjugates were separated from free GAs by gel permeation chromatography and then from each other by reversed-phase HPLC. Anion-exchange chromatography (DEAE-Sephadex A-25) on individual conjugates determined whether they were glucosyl ethers or esters. The latter were transesterified to methyl esters with NaOH/methanol. In addition the conjugates were hydrolysed with cellulase and the hydrolysis products converted to methoxycoumaryl esters. HPLC was carried out after each step and where possible an indication of identity was obtained by co-chromatography with authentic standards.

VI. METABOLISM STUDIES

Understanding of GA biosynthesis and catabolism has advanced considerably in recent years. These advances have come equally from work with fungal cultures, intact higher plants and cell-free preparations from higher plants (Bearder, 1983; Coolbaugh, 1983; Hedden, 1983; Sponsel, 1983). Almost all experiments designed to study metabolism have involved application of a substrate to a plant system and analysis of the resulting products. The analytical problems are similar to those discussed earlier in this chapter with the added difficulty that not only must the products be identified, but they must also be shown to originate from the substrate. This aspect is discussed more fully later in this section. Other considerations, such as the choice of suitable plant systems and the interpretation of results, are outside the scope of this chapter. However, it may be useful to point out the

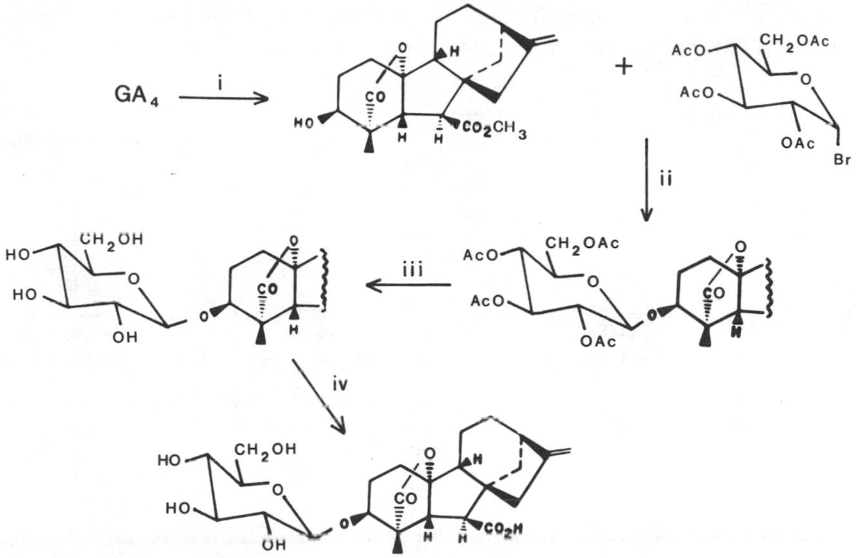

Fig. 25. Procedure for the synthesis of GA-glucosides (Schneider *et al.*, 1977). Reagents: i, CH_2N_2; ii, Ag_2CO_3; iii, CH_3ONa; iv, $LiSC_3H_7$.

difficulties in relating experimentally observed metabolic processes to those that occur naturally. Under experimental conditions unnatural reactions can be induced by applying an inappropriate substrate for the organ, tissue or developmental stage, or by applying excessive quantities of a suitable substrate. Furthermore, only a fraction of the applied substrate may be taken up by intact tissues, and of that very little may reach the site of metabolism. Much of the substrate may be diverted to sites such as the vacuole, where it is converted by non-specific enzymes to products that are not produced *in vivo*. Thus many intact plant tissues convert GAs predominantly to conjugates that are not found as endogenous components. The breakdown of cellular compartmentation produced as a result of homogenization may also lead to artefactual products. All these factors need to be considered when the results of such experiments are interpreted. Probably the best evidence that an experimentally observed pathway is natural is that both substrate and final products are endogenous to the tissue under study.

A. The use of isotopically-labelled compounds as tracers

Except in certain experiments with *G. fujikuroi* when endogenous GA biosynthesis was inhibited, substrates used in GA-metabolism studies have been labelled with one or more heavy isotopes. In most cases labelling is essential if the metabolic products are to be traced. The labels that can be used are the radio-isotopes ^{3}H and ^{14}C, and the stable isotopes ^{2}H, ^{13}C, ^{18}O and ^{17}O. The properties of these isotopes are listed in Table XII.

Table XII. Properties of relevant isotopes.

Isotope	Mass	Natural abundance (atom %)	Specific radioactivity (Bq/A)	Half life (yr)
^{1}H	1.007825	99.984	—	stable
^{2}H	2.01400	0.0156	—	stable
^{3}H	3.01605	—	1.07×10^{15}	12.35
^{12}C	12.00000	98.892	—	stable
^{13}C	13.00335	1.108	—	stable
^{14}C	14.0032	—	2.31×10^{12}	5730
^{16}O	15.99491	99.759	—	stable
^{17}O	16.99914	0.037	—	stable
^{18}O	17.99916	0.204	—	stable
^{28}Si	27.97693	92.21	—	stable
^{29}Si	28.97649	4.70	—	stable
^{30}Si	29.97376	3.09	—	stable

The radio-isotopes are favoured by most workers since they simplify the detection of products and the determination of recoveries and conversion efficiencies. The detection of metabolites labelled with stable isotopes is limited to MS and NMR. Table XIII lists the relative advantages and disadvantages of ^{3}H and ^{14}C as radiolabels.

Some clarification of the points in Table XIII is necessary. The choice of label will depend to a large extent on the application. It may be dictated by the methods available for introducing the isotope into a particular substrate (see Section VII) or by the amount of conversion anticipated. Tritiated compounds can be synthesized with specific radioactivities up to 1000 times higher than for those labelled with ^{14}C, and can thus be detected at much lower concentrations.

The isotope effect is due to the increased mass of the ^{3}H nucleus and results in reduced vibrational energy of the C-^{3}H bond compared to the C-^{1}H bond. Therefore C-^{3}H bonds, and to a lesser extent C-^{2}H bonds, require greater energy to break them than do C-^{1}H bonds. A primary isotope effect is observed in a reaction when the C-^{3}H (^{2}H) bond is broken as the rate-limiting step and results in the labelled substrate reacting more slowly than the unlabelled equivalent. Thus if a mixture of unlabelled and labelled substrates is used, which is generally the case with ^{3}H-labelled compounds, the unreacted substrate will contain increasing amounts of ^{3}H as the reaction proceeds. The product will have lost the label unless an intramolecular hydrogen transfer is involved. The point is illustrated in Fig. 26, which

Table XIII. Relative advantages and disadvantages of using ^3H and ^{14}C as radiolabels in metabolism studies.

Advantages	Disadvantages
^3H	
Simple to introduce into molecules Inexpensive, High specific radioactivity—detectable at low substrate concentration by liquid scintillation counting, Useful as a probe of reaction mechanisms.	Low energy emitter—detected inefficiently by most radiomonitors, Subject to exchange—may be lost non-enzymatically, Not usually detected by MS, labelled substrate may be converted more slowly than the natural substrate.
^{14}C	
Label usually more stable than ^3H, Efficiently detected by radiomonitors, Can be incorporated with high isotopic substitution—suitable for detection by MS, gives very small isotope effect—labelled substrates behave as unlabelled equivalent, chemically and biochemically.	More expensive and difficult to introduce than ^3H, Low specific radioactivity more labelled product required for detection.

compares the rates of oxidation of *ent*-7α-hydroxy [^{14}C]kaurenoic acid and *ent*-7α-hydroxy [6α-^3H]kaurenoic acid by resuspended microsomes from *C. maxima* endosperm (Graebe *et al.*, unpublished results). The products of the reaction, GA_{12}-aldehyde and *ent*-6α,7α-dihydroxykaurenoic acid (see Fig. 4), are found to contain no ^3H. The reaction involves breaking the *ent*-6α-H bond and shows an isotope effect of about 8. Secondary isotope effects, which are much smaller, involve the cleavage of bonds adjacent to that containing the isotope. Heavier atom isotopes such as ^{13}C and ^{14}C exhibit negligible isotope effect since the relative difference in masses is smaller, i.e. $^{14}C/^{12}C = 116\%$ as opposed to $^3H/^1H = 300\%$.

The loss of hydrogen isotopes from stereospecifically-labelled substrates can provide useful information on the mechanism of a reaction. In the conversion of *ent*-7α-hydroxykaurenoic acid to GA_{12}-aldehyde discussed above, the results illustrated in Fig. 26 indicate that the reaction is initiated by the loss of the *ent*-6α hydrogen atom and, because of the observed isotope effect, hydrogen abstraction is a rate-limiting step. In another example, Evans *et al.* (1970) studied the incorporation of specifically-labelled mevalonic acid (MVA) molecules into GAs by cultures of *G. fujikuroi*. The position of tritium atoms in the products were located by chemical degradation and measurement of $^3H/^{14}C$ ratios. They were able to show by feeding [2S-^3H,2-^{14}C]MVA and [5R-^3H,2-^{14}C]MVA that the 1α- and 2α-hydrogen atoms were lost in a *cis* elimination during formation of the 1,2 double bond in GA_3.

Specific labelling with heavy isotopes of carbon or oxygen has been used to examine aspects of mechanism not readily determined using ^2H or ^3H.

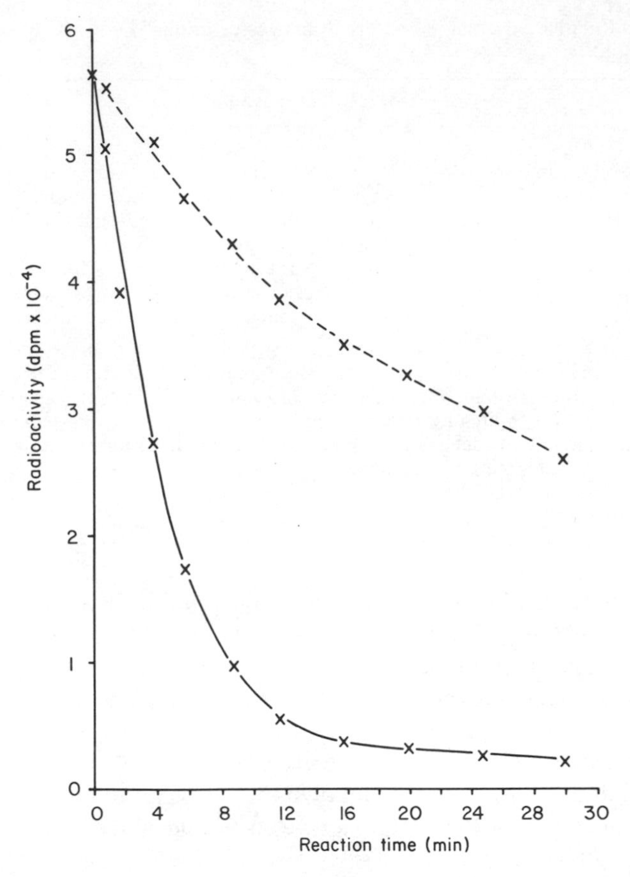

Fig. 26. Comparison of the rates of conversion of *ent*-7α-[^{14}C]kaurenoic acid (x——x) and *ent*-7α-hydroxy[6α-^{3}H]kaurenoic acid (x – – x) to GA_{12}-aldehyde and *ent*-6α,7α-dihydroxy-kaurenoic acid by a microsomal suspension from *Cucurbita maxima* endosperm (Graebe *et al.*, unpublished results).

Dockerill and Hanson (1978) produced *ent*-kaurene labelled *inter alia* at C-20 with ^{14}C. They followed the fate of this atom in the conversion to C_{19}-GAs in *G. fujikuroi* cultures and showed it was oxidized to CO_2. In combination with GC–MS, labelling with ^{18}O can be a valuable technique. Bearder *et al.* (1976b) showed that the γ-lactone in C_{19}-GAs contains both O atoms of the C-19 carboxyl group and were able to discount a Baeyer–Villiger reaction for the loss of C-20. Incubations of cell-free preparations with GA precursors in an atmosphere of $^{18}O_2/N_2$ have shown that oxygen atoms incorporated by hydroxylation reactions are derived from the air (Murphy and West, 1969; Hedden *et al.*, 1984).

(1) Detection of radio-isotopes

Radiochromatography (the detection of radioactively labelled compounds after chromatographic separation) involves several techniques that are adapted for the chromatographic system with which they are associated. In TLC the plate is moved relative to the collimator slit of a gas-flow proportional counter. The method is quick and non-destructive, and is useful as part of a purification procedure. Although ^{14}C can be detected with relatively high efficiency (15–30%), ^{3}H, which is a low-energy emitter, is detected poorly.

On-line detection of radiolabelled metabolites after GC (GC–RC) can also be accomplished with a gas-flow proportional counter. The GC effluent is split and a fraction is directed to a FID detector to provide a mass trace. The bulk of the effluent flows into a combustion chamber consisting of cupric oxide heated to about 700°C. Under these conditions carbon atoms are oxidized to carbon dioxide, which is swept into the counting tube by the carrier gas (usually argon). Tritiated compounds are reduced to hydrogen by passing over heated iron wire in the presence of hydrogen. Three dual traces of the same sample after GC on different columns are shown in Fig. 27.

An alternative inexpensive technique utilizes the FID detector as the combustion chamber (Wels, 1977). The FID effluent, $[^{3}H]H_2O$ or $[^{14}C]CO_2$, is trapped, mixed with scintillation fluid and counted in a liquid scintillation counter. The resolution of the output depends on the size of the fractions collected. When tritiated samples are detected the water produced in the FID is condensed and either dripped into scintillation vials or first mixed on-stream with the scintillation fluid before dripping into the vials. The recovery of tritium is very high, so the method can be used for quantitation. For trapping of $[^{14}C]CO_2$, a base such as ethanolamine is included in the scintillation fluid. Efficient mixing of the FID effluent with the scintillation stream is essential for high recoveries of counts.

Continuous-flow radioactivity monitoring of HPLC eluents is now common practice, and numerous commercial HPLC radiomonitors are available. Reeve and Crozier (1977) discussed the use of a homogeneous system in which scintillation fluid was mixed on-line with the HPLC effluent and the mixture passed through a counting cell positioned between two photomultiplier tubes. Counting efficiencies were high for both ^{14}C (60–80%) and ^{3}H (8–35%) depending on the HPLC solvent. Heterogeneous counting systems are also available. In this case the counting cell is packed with a solid scintillant such as cerium-activated glass or yttrium silicate-coated glass. These systems are non destructive and simple to use. However, counting efficiencies are less than those obtained with the homogeneous system and the solid scintillant is prone to contamination. Figure 28 illustrates an HPLC

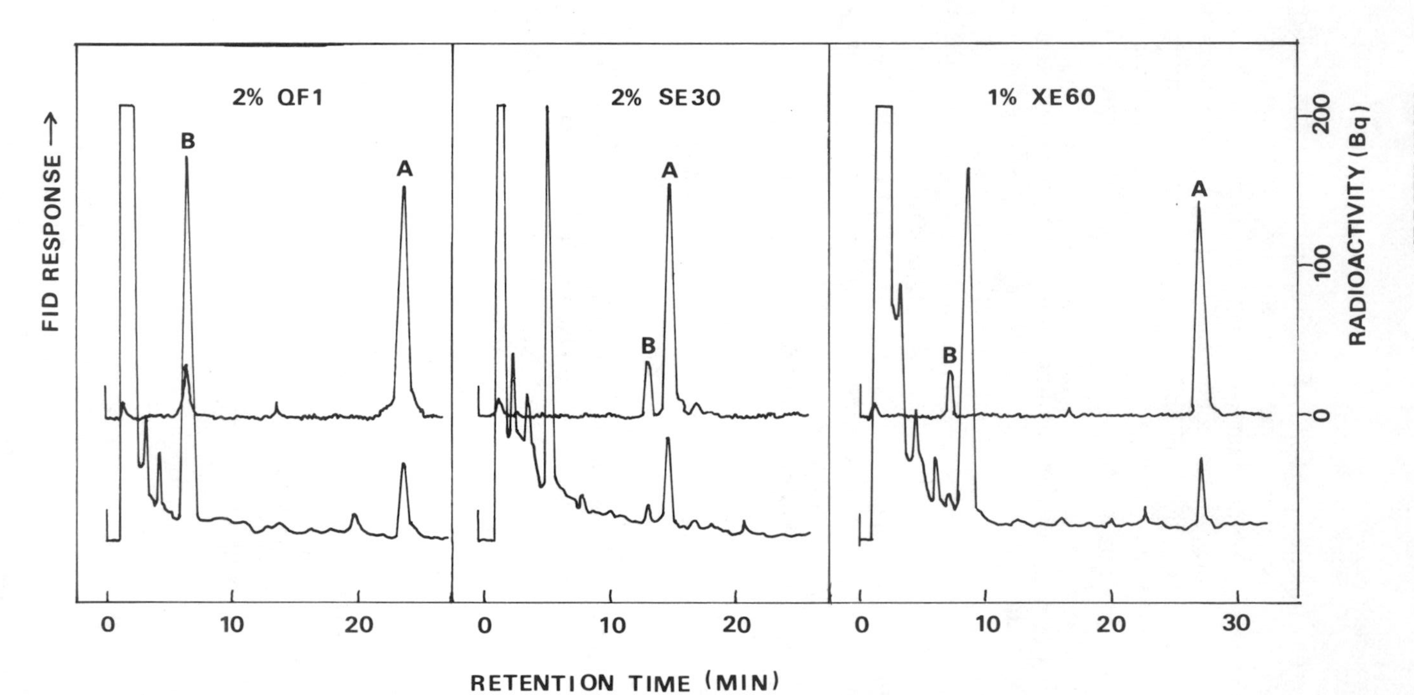

Fig. 27. GC-FID and RC traces for a MeTMS derivatized fraction from an extract of *Pisum sativum* seed after application of [2,3-$^3H_2$]GA_{20} (Durley *et al.*, 1979). The fraction contained predominantly two labelled products, a GA_{29} catabolite (A) and a minor component, "metabolite B". The GC columns (1.8 × 2 mm) were maintained at 206°C (2% QF-1), 203 (SE-30) and 209°C (XE-60). A helium gas flow rate of 55 ml min_{-1} was used. The RC trace was obtained using a modified Nuclear-Chicago 4998 gas flow proportional counter. The traces were kindly provided by Dr R. C. Durley.

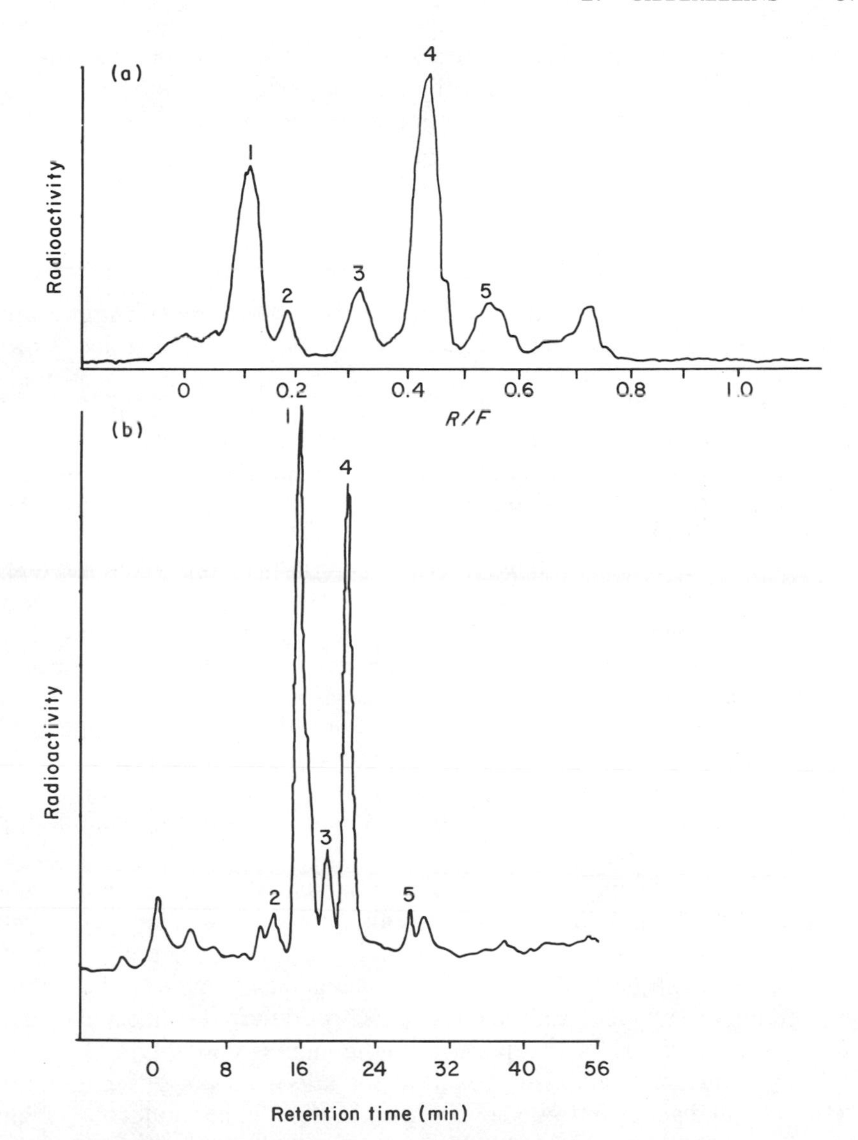

Fig. 28. (a). TLC radiochromatogram of an aliquot of the total extract from an incubation of 3*R*[2-^{14}C]MVA with a cell-free system from *Cucurbia maxima* endosperm. The 0.25 mm silica gel plate was eluted with chloroform/ethyl acetate/acetic acid (75:25:1). Radioactivity was monitored using a Panax-Nucleonics E0111/7973 plate scanner. (b). The same extract separated by reversed-phase HPLC on a 25 × 0.5 cm Partisil-10 ODS-3 column eluted with a gradient from 50% methanol/water to 100% methanol over 50 min and then isocratically. Solvents contained 50 $\mu l\,l^{-1}$ acetic acid. Radioactivity was detected using a Berthold LB503 HPLC radioactivity monitor. Peak assignments: 1, *ent*-6α, 7α-dihydroxykaurenoic acid; 2, GA_{12}; 3, *ent*-7α-hydroxykaurenoic acid; 4, GA_{12}-aldehyde; 5, *ent*-kaurenol and *ent*-kaurenoic acid.

radiochromatogram trace of the products from an incubation of [2-^{14}C]MVA with a cell-free system from *C. maxima* endosperm. For comparison, a TLC radiochromatogram of the same product mixture is also shown.

B. Validity of product identity

It is important not only that metabolic products be adequately identified, but also that they be shown to have originated from the applied substrate. When radioactive isotopes are used as tracers it is common practice to identify the products by comparing their chromatographic properties with those of standards. The limitation of this method is the same as that discussed for qualitative analysis (Section IV.A). The discriminating power of chromatographic methods is usually insufficient for identification to be based solely on co-chromatography. Gas chromatography provides the highest resolution, and identity has been based on GC rentention times on three columns with different polarity phases (see Fig. 27) (Durley *et al.*, 1979). In practice this method has proved quite reliable since the number of possible identities of the radiolabelled products is severely restricted. Co-crystallization of radioactive products with authentic standards has been used extensively as a basis for identity. This method, however, does not distinguish compounds that co-crystallize. Such compounds are likely to be structurally similar and may not separate during sample purification. Co-crystallization requires also that the identity of the product is suspected and that relatively large amounts of standard are available.

In many cases the identity of the product is unknown and/or standards are not available for comparison. It is also seldom the case that enough product can be isolated for a chemical characterization. MS is then the only method available for identification. However, for labelled compounds to be distinguished by MS sufficient isotope must be incorporated to be detected by the mass spectrometer. High incorporation of the stable isotopes ^{2}H, ^{13}C, ^{17}O or ^{18}O is possible, but radio-isotopes have the added benefit of being more easily detected and quantitated during purification. Since tritium has a high specific radioactivity (1.07×10^{15} Bq A^{-1}) it is seldom practicable to use tritiated substrates with an isotopic substitution sufficient for detection by MS. Sponsel and MacMillan overcame this problem by using a combination of ^{2}H and ^{3}H substituted at the same position in the molecule. As well as being able to show convincingly the identity of the metabolic products, they could estimate the concentrations of the endogenous metabolites. As an example, Sponsel and MacMillan (1977) fed a mixture of [1β, 3α-$^{2}H_2$]- and [1β, 3α-$^{3}H_2$]GA_{20} (1.4 mg, containing 1.67 atoms ^{2}H molecule^{-1} and

4.03×10^4 Bq) to immature *P. sativum* seeds. After 4 days the products were separated by TLC. A polar zone containing 78% of the radioactivity was found by MS to contain GA_{29} with a strong $[M+2]^{+\cdot}$ ion associated with the molecular ion (estimated ^2H content was 0.8 atoms molecule^{-1}). Thus GA_{29} was formed from the exogenous GA_{20} in 78% yield, i.e. 1.15 mg GA_{29} were produced from the applied substrate. The ratio of exogenous to endogenous GA_{29} can be calculated as 0.8/(1.67–0.8):1 = 0.92:1. Therefore the weight of endogenous GA_{29} in the seeds at the time of extraction was 1.15/0.92 = 1.25 mg. Since GA_{20} was also recovered from the seeds its concentration could be similarly calculated.

The relatively low specific radioactivity of ^{14}C (2.33×10^{12} Bq A^{-1}) enables it to be used as both a radiotracer and a heavy isotope for MS detection. Molecules with a specific radioactivity of 7×10^{11} Bq mol^{-1} or greater containing a single ^{14}C atom can be detected easily by MS. Multiple substitution with ^{14}C allows simpler detection of the label even when the product is heavily diluted with endogenous compound (Hedden and Graebe, 1981). High-specific-radioactivity [2-^{14}C]MVA is incorporated with four ^{14}C atoms into *ent*-kaurenoids and GA products, which give $[M+8)^{+\cdot}$ ions that are easily distinguished by MS. Bowen *et al.* (1972) calculated the specific radioactivities of ^{14}C-labelled products from the relative intensities of the isotope peaks in the molecular ion clusters. Thus, as for the [^2H, ^3H]-labelled substrates, in combination with radio-counting the concentrations of endogenous metabolites can be calculated.

VII. PREPARATION OF ISOTOPICALLY-LABELLED GIBBERELLINS AND GIBBERELLIN PRECURSORS

Numerous chemical methods have been employed for labelling GAs and related compounds. In addition, biological systems, such as fungal cultures and cell-free enzyme preparations, in combination with chemical synthesis, have provided routes to labelled compounds that could otherwise be prepared only with difficulty.

A. Chemical methods

(1) Catalytic hydrogenation

Tritiated GAs of high specific radioactivity can be prepared by catalytic reduction with carrier-free tritium. [1,2-$^3H_2$]GA_1 was prepared from GA_3 in about 40% yield using a palladium catalyst that was partially poisoned to inhibit hydrogenation of the 16,17 double bond (Kende, 1967; Pitel and Vining, 1970; Nadeau and Rappaport, 1974). Nadeau and Rappaport, using palladium on calcium carbonate poisoned with pyridine, prepared tritiated GA, of specific radioactivity 1.59×10^{15} Bq mol^{-1} in 37% yield. Other products included 16,17 dihydro derivatives (5%), dibasic acids formed by hydrogenolysis of the γ-lactone (26%) and, surprisingly, GA_3 (9.4%). The latter had a specific radioactivity of 4.8×10^{14} Bq mol^{-1} and must have been produced by exchange of alkane hydrogen atoms. Nadeau and Rappaport assumed that the [^3H]GA_1 was also randomly substituted to the extent of 4.8×10^{14} Bq mol^{-1} and thus labelled at the 1 and 2 positions to the extent of 1.11×10^{15} Bq mol^{-1} (maximum theoretical value is 2.14×10^{15} Bq mol^{-1}). Dilution of label must have occurred by exchange with the solvent (THF). In order to minimize exchange it is important to use aprotic solvents. [1,2-$^3H_2$]GA_4 has been prepared from GA_7 in an analogous manner (Durley and Pharis, 1973). [1,2-$^3H_2$]GA_1, and [1,2-$^3H_2$]GA_4 are now available commercially from Amersham International plc.

The 16,17-double bond can be protected during hydrogenation by forming the epoxide with *m*-chlorobenzoic acid. The double bond is restored later by reduction with zinc, sodium iodide and sodium acetate. Murofushi *et al.* (1977) employed this procedure to prepare [2,3-$^3H_2$]GA_{20} from GA_5, as did Yokota *et al.* (1976), who reduced 2,3-dehydro GA_9, prepared from GA_4 via the methane-sulphonate, to [2,3-$^3H_2$]GA_9.

(2) Exchange reactions

GA_3 has been labelled randomly with tritium by exchange using the Wilzbach method (Baumgartner *et al.*, 1959; Murofushi *et al.*, 1977). However, the product has a low specific radioactivity and much of the tritium is readily lost. Ayrey and Chapman (1979) described a platinum oxide-catalysed exchange with tritiated water (1.85×10^{12} Bq mol^{-1}) to produce [^3H]GA_3 of high specific radioactivity (2.47×10^{14} Bq mol^{-1}), but in very low yield (2.7%). The position of the label was not determined and is assumed to be random. The yield and specific radioactivity were lower than those obtained by Nadeau and Rappaport (1974) when GA_3 was produced as a by-product.

Acid- or base-exchangeable hydrogen atoms can be conveniently replaced with ^2H or ^3H. The labelled products are usually stable at non-extreme pH values or, better, can be stabilized by further reactions. In the absence of a 13-hydroxyl group the 16,17 double bond is isomerized by strong acid to give a mixture of 15,16 and 16,17 double-bond isomers (Fig. 29). Bearder *et al.* (1974) used $CF_3CO_2{}^3H(^2H)$ to label *ent*-kaurene at C-15

Fig. 29. Reversible, acid-catalysed isomerization of the 16,17 double bond to the 15,16 position. In the presence of tritiated acid such as $CF_3CO_2{}^3H$, tritium is introduced at C-15 and C-17.

and C-17. The double-bond isomers were separated by TLC on $AgNO_3$-impregnated silica gel.

Base-catalysed exchange of hydrogen atoms situated α to carbonyl groups can be exploited. GA_{12}-aldehyde and GA_{14}-aldehyde were labelled at C-6 by treatment with HO^3H (^2H) or $CH_3O\ ^3H(^2H)$ and CH_3ONa (Bearder *et al.*, 1973; Hedden *et al.*, 1974) (Fig. 30). Lischewski *et al.* (1982) utilized this reaction for the preparation of $[6\text{-}^3H_1]GA_3$. GA_3-7-aldehyde was prepared via the anhydride (Lischewski and Adam, 1980). After protection of the hydroxyl groups by trimethylsilylation, base exchange was conducted in THF with tritiated water and potassium hydride. The resulting mixture of $[6\text{-}^3H]GA_3$-7-aldehyde and its 6-epimer was resolved by TLC. Oxidation of the 7-aldehyde gave $[6\text{-}^3H]GA_3$ of specific radioactivity 3.88×10^{12} Bq mol^{-1}.

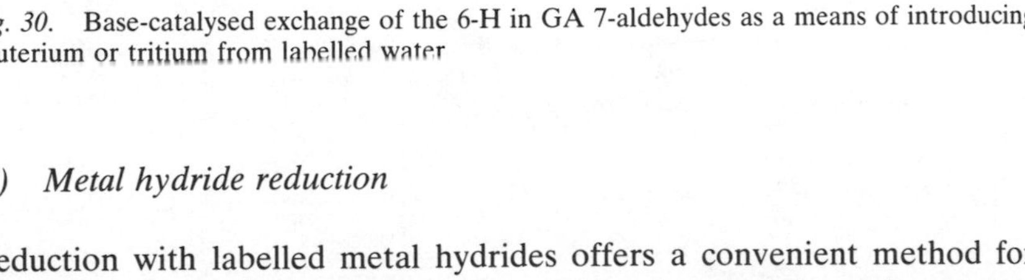

GA_{12} - aldehyde R = H
GA_{14} - aldehyde R = OH

Fig. 30. Base-catalysed exchange of the 6-H in GA 7-aldehydes as a means of introducing deuterium or tritium from labelled water

(3) *Metal hydride reduction*

Reduction with labelled metal hydrides offers a convenient method for introducing ^2H or ^3H (see MacMillan, 1980). An example of the preparation of several labelled GAs from GA_3 or GA_7 is illustrated in Fig. 31. Beale and MacMillan (1980) described the reduction of the GA_3 and GA_7 enone derivatives (**13** and **14**) with BH_4^- in an aprotic solvent. The major products are the 3α-alcohols (**15** and **16**) in which hydrogen from BH_4^- is introduced at the 1β and 3β positions. Small amounts of the 3β-alcohols (**17** and **18**) are also formed. A third hydrogen is introduced at C-2 from the

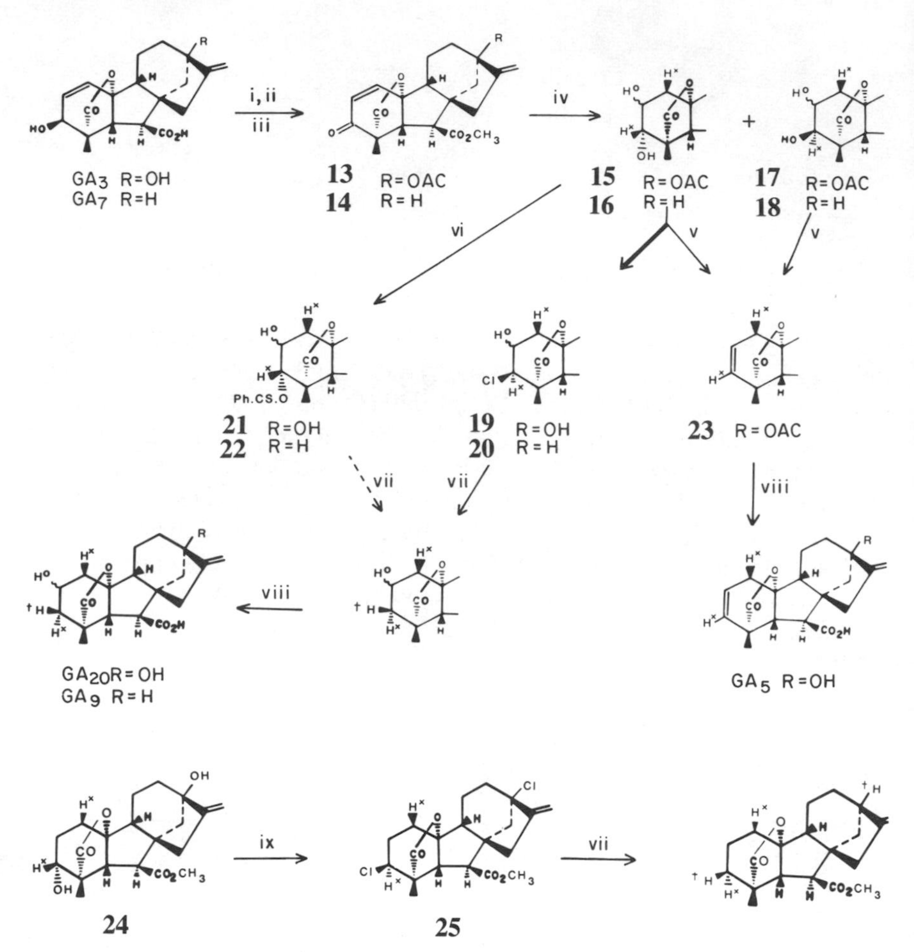

Fig. 31. Scheme for introducing deuterium or tritium into GAs by metal hydride reduction (Beale and MacMillan, 1980; Beale *et al.*, 1980; Duri *et al.*, 1981). Reagents (symbols $H^{\times}$, H^{o}, $^{\dagger}H$ are used to trace the origins of the hydrogen atoms that are introduced): i, CH_2N_2; ii, MnO_2; iii, $(CH_3CO)_2O$, p-$CH_3C_6H_4SO_3H$; iv, $NaBH^{\times}_4$, LiBr–H^{o}_2O, $H^{o}Cl$ used in the work-up; v, $POCl_3$, C_5H_5N; vi, $C_6H_5C(Cl) = N^+(CH_3)_2Cl^-$, H_2S; vii, $(n\text{-}C_4H_9)_3Sn^{\dagger}H$; viii, KOH; ix, $(C_6H_5)_3P$, CCl_4, C_5H_5N. The discontinuous arrow denotes that the configuration of the label in the product is unknown.

protic solvent during the work-up. Thus there is an opportunity to label with up to three ^2H or ^3H atoms or a mixture of isotopes.

The Me ester of 3-epi-GA_4 (**16**) can be converted to GA_4Me (**18**) in low yield by base-catalysed epimerization (Beale *et al.*, 1980). The label at C-2 is then lost by exchange. Alternatively, oxidation to the 3-ketone and reduction with aluminium tri-isopropoxide gives GA_4Me in better yield labelled at C-1 and C-2 (Beale and MacMillan, 1980). High yields of GA_1 or GA_4 from the 3-ketones can be obtained using potassium tri-sec-butyl-borohydride (K-selectride) as reductant and anhydrous potassium hydrogen phosphate as a proton source (Bell and Turner, 1981). The 3-epi-GA_1 and GA_4 derivatives (**15** and **16**) were converted to GA_{20} and GA_9 respectively without loss of label by first forming either the 3β-chloro derivatives (**19** and **20**) or the 3β-thiobenzoates (**21** and **22**), which were reduced with tri-n-butylstannane (Beale *et al.*, 1980). ^3H or ^2H could be introduced via labelled stannane at this stage. The reaction of phosphorus oxychloride with the 3-alcohols (**15** and **17**) caused some dehydration to give GA_3 Me acetate (**23**). The free alcohols are obtained from the Me ester 13-acetates by alkaline hydrolysis.

Duri *et al.* (1981) converted the 3-epi GA_1 derivative (**24**) into the 3β,13-dichloro derivatives (**25**) with triphenylphosphine and carbon tetrachloride. Substitution of chloride with hydride from tri-n-butylstannane allowed for further incorporation of label.

(4) Methylenation

The replacement of C-17 offers the only simple method to introduce a carbon label, and can also be used to introduce hydrogen isotopes that are not readily exchangeable. The 17-norketones are synthesized by reaction of sodium periodate and a catalytic quantity of osmium tetroxide (Fig. 32) or by ozonolysis. A labelled methylene group can then be introduced by the Wittig reaction. The reactive species (an ylide) is produced by the action of strong base on a methyltriphenylphosphonium salt (Fig. 32). Bearder *et al.* (1976a) prepared the ylide from methyltriphenylphosphonium bromide in

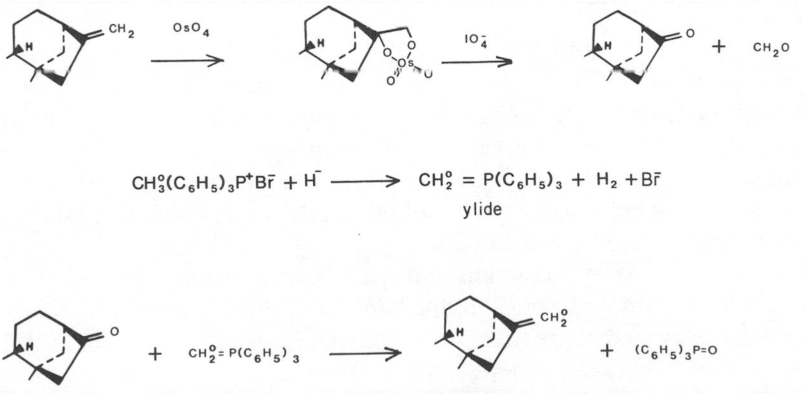

CH_2 $\xrightarrow{OsO_4}$ (osmate ester) $\xrightarrow{IO_4^-}$ =O + CH_2O

$$CH_3^{\circ}(C_6H_5)_3P^+Br^- + H^- \longrightarrow CH_2^{\circ} = P(C_6H_5)_3 + H_2 + Br^-$$

ylide

=O + $CH_2^{\circ}{=}P(C_6H_5)_3$ $\longrightarrow$ =CH_2° + $(C_6H_5)_3P{=}O$

Fig. 32. Introduction of label at C-17 by use of the Wittig reaction. In this example the ylide is generated from the methyltriphenylphosphonium bromide using a hydride as base.

dry THF (redistilled from calcium hydride) using either potassium t-butoxide or sodium hydride as base. In the latter case an essentially salt-free solution of the ylide is generated since the by-product, sodium bromide, precipitates. The solution can be stored in a sealed container and portions used as required.

Labelled methyltriphenylphosphonium salts can be purchased or synthesized by the reaction of methyl halide with triphenylphosphine. Bearder *et al.* (1976a) prepared [C_1 $^3H_3$]triphenylphosphonium bromide by exchange with tritiated water under mildly basic conditions. The reaction of the ylide with the ketone is carried out in dry THF under nitrogen. Since the reaction conditions are extremely basic, base-labile functional groups must be protected. Thus 3β-hydroxy GAs such as GA_4 can be protected as TMS ethers, which are stable under the anhydrous reaction conditions. The basic conditions also cause scrambling of the hydrogen atoms between C-17 and C-15 in 13-deoxy GAs and between C-17, C-15 and C-14 in 13-hydroxy GAs (Bearder *et al.*, 1976a).

Although *ent*-kaurenoids and non-polar GAs can be labelled in reasonably high yield, more highly-functionalized GAs are prone to side reactions and often give poor yields. An alternative to the Wittig reaction has been reported recently to give much better yields (Lombardo, 1982). The reagent, which is of unknown structure, is generated from zinc dust, dibromomethane and titanium tetrachloride. Incorporation of deuterium from [$^2H_2$]CH_2Br_2 was reported to be completely specific to C-17.

B. Biochemical methods

(1) Fungal cultures

Certain strains of the fungus, *G. fujikuroi* (imperfect form, *Fusarium moniliforme*) are used for the commercial production of GA_3 and other GAs. This fungus may also be used to produce labelled GAs by incubation with labelled precursors. Thus [^{14}C]GA_3 can be produced from [^{14}C]MVA, although the specific radioactivity of the product is low due to dilution with endogenous GA. Hanson and Hawker (1973) synthesized [17-^{14}C]GA_{12}-7-alcohol by the Wittig reaction and used fungal cultures to convert it to [17-^{14}C]GA_3, but with considerable dilution of the label. In order to obtain higher specific radioactivities, endogenous GA synthesis can be reduced. The use of the GA-deficient mutant, B1-41a, was discussed in Section IV.A.2. Since the mutant is blocked for GA biosynthesis before *ent*-kaurenoic acid, this substrate or later precursors are converted to the fungal GAs (Bearder *et al.*, 1975b) with little dilution of the label. Alternatively,

GA synthesis can be blocked by inhibitors. Several inhibitors such as the quaternary salts (Fig. 33a) prevent the synthesis of *ent*-kaurene without affecting the growth of the fungus. A second class of inhibitors that inhibit the oxidation of *ent*-kaurene to *ent*-kaurenoic acid has been discovered recently (Fig. 33b). These compounds are more effective than the quaternary salts but have fungicidal activity at high concentrations (Coolbaugh *et al.*, 1982).

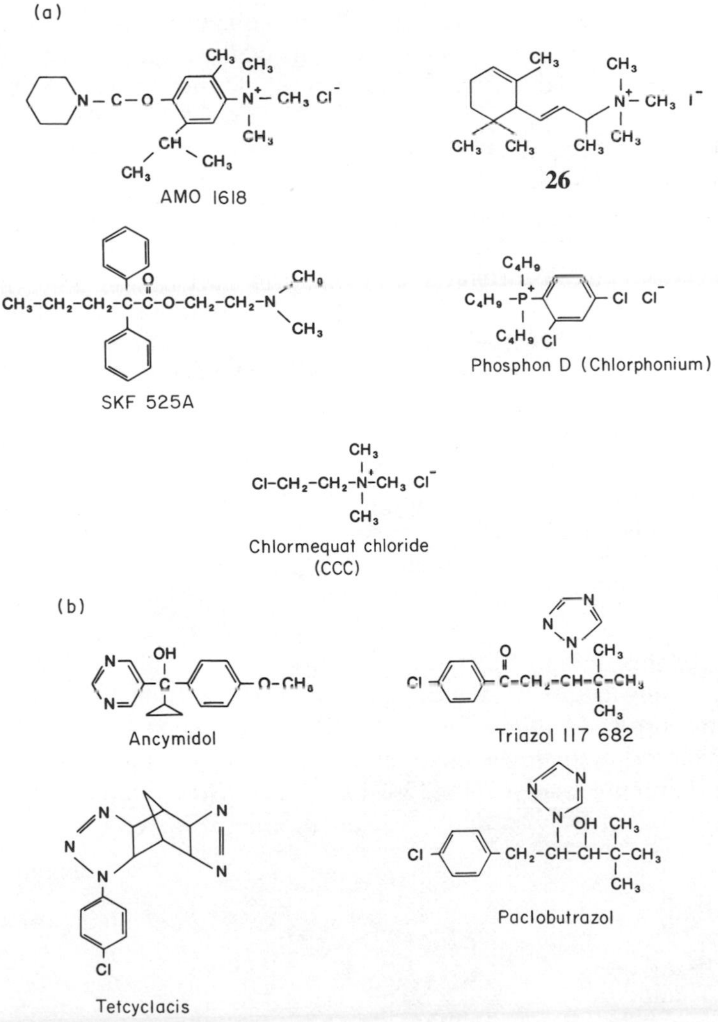

(a)

AMO 1618

26

SKF 525A

Phosphon D (Chlorphonium)

Chlormequat chloride (CCC)

(b)

Ancymidol

Triazol 117 682

Paclobutrazol

Tetcyclacis

Fig. 33. Inhibitors of GA biosynthesis. Group a inhibit the synthesis of *ent*-kaurene and Group b inhibit its conversion.

G. fujikuroi produces GAs as secondary metabolites when in the idiophase of growth, which is reached when a nutrient has been depleted. High GA production is achieved by use of a growth medium in which the carbon/nitrogen ratio is carefully chosen to ensure strong initial mycelial growth, but eventual depletion of nitrogen. Synthetic media based on the ICI medium (Borrow *et al.*, 1961), but containing only *ca.* 40% of the ammonium nitrate (40% ICI), are commonly used. The onset of the idiophase, which is indicated by pigmentation, is the stage at which substrates can be added. For minimum endogenous GA synthesis, inhibitors need to be present in the growth medium from the time of inoculation. Alternatively, a mycelium resuspension or replacement culture can be used. In this case, after pigmentation has started the mycelia are separated from the medium by filtration, washed and then resuspended in fresh nitrogen-free medium or buffer containing the substrate and inhibitor. This manipulation removes endogenous GAs, but not non-polar precursors such as *ent*-kaurene. Renewed GA synthesis should be prevented by the inhibitor.

The pattern of GAs produced by *G. fujikuroi* depends on the strain and growing conditions. Low pH favours the production of GA_3, but at higher pH intermediates accumulate (Bearder *et al.*, 1975b). A second GA-producing fungus, *Sphaceloma manihoticola,* produces GA_4, but not GA_7, GA_3 or GA_1 (Rademacher and Graebe, 1979; Ziegler *et al.*, 1980). It is a potential source of GA_4 uncontaminated by GA_7 from which it is difficult to separate.

Probably the most useful feature of *G. fujikuroi* is its ability to convert certain analogues of GA precursors to the corresponding GA-analogues (Hedden *et al.*, 1978). Thus it is possible to produce non-fungal GAs that are otherwise all but impossible to obtain in substantial quantities. Since the products are not endogenous, label introduced via the substrate is undiluted. However, in order to simplify purification of the products and to increase conversion by reducing substrate competition, the use of an inhibitor is still recommended.

An example of analogue metabolism is the conversion of steviol (*ent*-13-hydroxykaurenoic acid) (**27**) to 13-hydroxy GAs. Steviol occurs naturally in

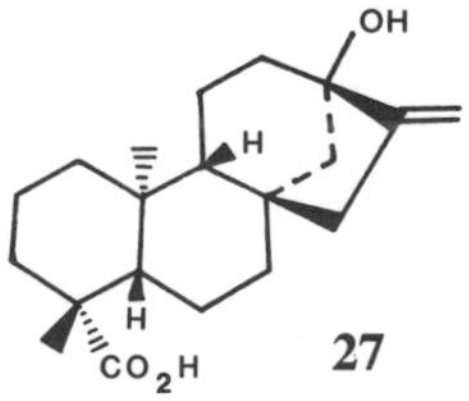

27

Stevia rebaudiana as the glycoside, stevioside, from which it can be prepared by enzymatic hydrolysis (Bearder *et al.*, 1975c). Table XIV lists the products with yields obtained from incubations of steviol with B1-41a, as reported by Bearder *et al.* (1975c). Also shown are products from incubations with steviol acetate. Since acetylation of the 13-hydroxyl group causes inhibition of 3β-hydroxylation, steviol acetate is converted to 3-deoxy products such as GA_{20} and GA_{17} acetates. By this method Bearder *et al.* (1976a) prepared [^{3}H]GA_{20} from [^{3}H]steviol acetate, which they had labelled via a Wittig reaction. Murofushi *et al.* (1979) used cultures of *G. fujikuroi*, strain G-2, grown in the presence of the inhibitor **26** (Fig. 33A), to prepare GA_1, GA_{18}, GA_{19} and GA_{53} from steviol. In an analogous manner, Gianfagna *et al.* (1983) prepared labelled GAs and *ent*-kaurenoid compounds from [17-^{14}C]- and [15,17-$^2H_2$]steviol in cultures of *G. fujikuroi*, strain LM-45-399, containing CCC. The yields of GAs were very low. Further examples of the microbiological production of non-fungal GAs were discussed in Section IV.A.2. In each case, since the substrates are easily labelled via the Wittig reaction or otherwise, there is clearly the potential for preparing many uncommon GAs in labelled form.

Table XIV. The yields of GAs produced from incubations of steviol and steviol acetate with *G. fujikuroi*, mutant B1-41a.

	Substrate		
Product	Steviol[a]	% Yield	Steviol acetate[b]
GA_1	26		—
GA_{18}	15.5		—
GA_{19}	3.5		—
GA_{53}	18.5		—
GA_{17} acetate	—		4
GA_{20} acetate	—		20
ent-kaurenoids	31.5		10

[a]Steviol incubated at 2 mg per 10 ml resuspended mycelium for 5 days (Bearder *et al.*, 1975c).
[b]Steviol acetate incubated at 9.37 mg per 100 ml resuspended mycelium for 7 days (Bearder *et al.*, 1976a).

(2) *Cell-free preparations*

Endosperm homogenates from the cucurbits, *Marah macrocarpa* and *C. maxima*, contain high enzyme activity for the conversion of MVA to GAs and GA precursors (West, 1973; Graebe, 1982). (3*R*)-MVA is converted into *ent*-kaurene and later *ent*-kaurenoids in virtually quantitative yield. Under suitable conditions the most active preparations from *C. maxima* produce GA_{12}-aldehyde and *ent*-$6\alpha,7\alpha$-dihydroxykaurenoic acid as the major products (Graebe *et al.*, 1972). The former intermediate is further converted to several C_{20}-GAs (Graebe *et al.*, 1974).

Although the scale of the conversions is small, usually in the μg range, the incubations produce intermediates of very high specific radioactivity, which are valuable for metabolism studies. Typically, [2-^{14}C]MVA of specific activity 1.96×10^{12} Bq mol^{-1} is converted by the *C. maxima* preparation to [^{14}C]GA_{12}-aldehyde of specific activity 3–4.5×10^{12} Bq mol^{-1}. Some dilution of label occurs due to the presence of *ent*-kaurene in the microsomes. Since *ent*-kaurene is biosynthesized from MVA by soluble enzymes, it can be produced without dilution if the microsomes are removed from the preparation by high-speed centrifugation. The position of the label that is incorporated into *ent*-kaurene and GA_{12} from [2-^{14}C]MVA is indicated in Fig. 34.

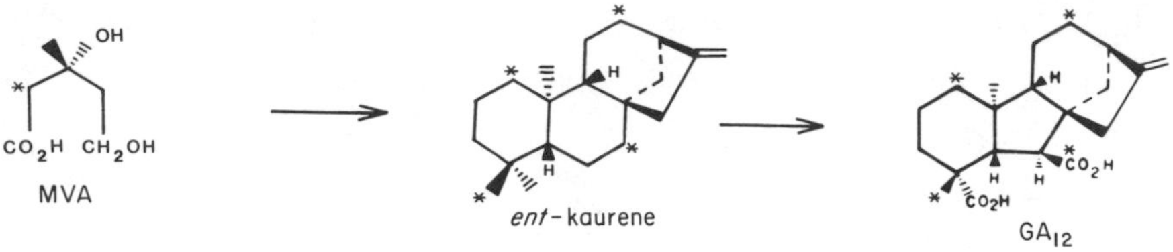

Fig. 34. Position of the label in *ent*-kaurene and GA_{12} produced enzymatically from [2-^{14}C]MVA.

Enzyme activities from different *C. maxima* fruit are variable. Highest activity is obtained from endosperm in seeds in which the cotyledon length is 20–70% the length of the lumen. The limiting factor is the activity of the microsomal mixed-function oxygenases that convert *ent*-kaurene to GA_{12}-aldehyde. It is often necessary to screen several fruit before an active preparation is found. The endosperm from each fruit should be processed and stored separately to avoid diluting active preparations with poorer ones. The preparation of the cell-free system has been described in detail (Graebe *et al.*, 1974). A low-speed ($2000 \times$ g) supernatant fraction is incubated with MVA in the presence of ATP, PEP, magnesium chloride, manganese chloride and NADPH for about two hours at 30°C. The biosynthesis can be stopped at *ent*-kaurene simply by omitting NADPH from the incubation. The conversion of GA_{12}-aldehyde or GA_{12} to C_{20}-GAs is catalysed by the high-speed (150 000–200 000 $\times$ g) supernatant and requires Fe^{2+} as sole exogenous cofactor (Hedden and Graebe, 1982). Manganese chloride inhibits these later steps and its presence in the incubation mixture ensures that the biosynthesis from MVA stops at GA_{12}-aldehyde and GA_{12}. The end-product of the pathway is GA_{43}, but intermediates accumulate if the ratio of substrate to cell-free preparation is high. The conditions necessary to obtain a particular intermediate are largely empirical and vary from one preparation to another.

In vitro enzyme preparations from other sources have been used to prepare particular intermediates (see Hedden, 1983). One of the more useful systems for preparing 13-hydroxy GAs was obtained from *P. sativum* seeds. Kamiya and Graebe (1983) developed a system from the cotyledons that converted GA_{12} to several 13-hydroxy GAs including GA_{19}, which is an important intermediate in higher plants and is difficult to prepare in labelled form by any other means. These workers prepared the substrate, $[^{14}C]GA_{12}$, from $[2\text{-}^{14}C]MVA$ with the *C. maxima* system.

Future progress in GA biochemistry will depend to a large extent on the availability of appropriately-labelled compounds. Combinations of chemical, microbiological and *in vitro* enzyme methods such as those described above can be potentially of great value for the preparation of a wide range of scarce GAs and precursors in labelled form. The scope of the methods is limited only by the ingenuity and determination of the investigator.

ACKNOWLEDGEMENTS

The author wishes to acknowledge the help of several members of staff at East Malling Research Station. Mrs K. J. Carter produced some of the GC–MS data, Mrs K. Webster produced many of the figures and Miss P. A. Feltham typed the manuscript. Finally I thank many colleagues for their helpful criticism of the manuscript.

REFERENCES

Andersen, R. A. and Sowers, J. A. (1968). Optimum conditions for binding of plant phenols to insoluble polyvinylpyrrolidone. *Phytochemistry* **7**, 293–301.

Andersson, B. and Andersson, K. (1982). Use of Amberlite XAD-7 as a concentrator column in the analysis of endogenous plant growth hormones. *J. Chromatogr.* **242**, 353–358.

Atzorn, R. and Weiler, E. W. (1983a). The immunoassay of gibberellins 1. Radioimmunoassays for the gibberellins A_1, A_3, A_4, A_7, A_9 and A_{20}. *Planta* **159**, 1–6.

Atzorn, R. and Weiler, E. W. (1983b). The immunoassay of gibberellins 2. Quantitation of GA_3, GA_4 and GA_7 by ultra-sensitive solid-phase enzyme immunoassays. *Planta* **159**, 7–11.

Ayrey, G. and Chapman, J. M. (1979). Preparation of tritium-labelled gibberellin GA_3 at high specific activity. *J. Labelled Compd. Radiopharm.* **16**, 887–890.

Barendse, G. W. M., Van de Werken, P. H. and Takahashi, N. (1980). High performance liquid chromatography of gibberellins. *J. Chromatogr.* **198**, 449–455.

Baumgartner, W. E., Lazer, L. S., Dalziel, A. M., Cardinal, E. V. and Varner, E. L. (1959). Determination of gibberellins by derivative labeling with diazomethane-C^{14} and by isotope dilution analysis with tritium-labeled gibberellin. *Agric. Food Chem.* **7**, 422–425.

Beale, M. H. and MacMillan, J. (1980). Mechanism and stereochemistry of conjugate reduction of enones from gibberellins A_3 and A_7. *J. Chem. Soc. Perkin Trans.* I, 877–884.

Beale, M. H., Gaskin, P., Kirkwood, P. S. and MacMillan, J. (1980). Partial synthesis of gibberellin A_9 and [3α- and 3β-$^2H_1$]gibberellin A_9; gibberellin A_5 and [1β,3-$^2H_2$ and -$^3H_2$]gibberellin A_5; and gibberellin A_{20} and [1β,3α-$^2H_2$ and -$^3H_2$]gibberellin A_{20}. *J. Chem. Soc. Perkin Trans.* I, 885–891.

Beale, M. H., Bearder, J. R., Hedden, P., Graebe, J. E. and MacMillan, J. (1984). Gibberellin A_{58} and *ent*-6α,7α,12α-trihydroxykaur-16-en-19-oic acid from seeds of *Cucurbita maxima. Phytochemistry* **23**, 565–567.

Bearder, J. R. (1983). *In vivo* diterpenoid biosynthesis in *Gibberella fujikuroi:* The pathway after *ent*-kaurene. *In* "The Biochemistry and Physiology of Gibberellins" (A. Crozier, ed.) Vol. I, pp. 251–387. Praeger, New York.

Bearder, J. R. and MacMillan, J. (1980). Separation of gibberellins and related compounds by droplet counter-current chromatography. *In* "Gibberellins—Chemistry, Physiology and Use" (J. R. Lenton, ed.) pp. 25–30. British Plant Growth Regulator Group, Wantage.

Bearder, J. R., MacMillan, J. and Phinney, B. O. (1973). 3-Hydroxylation of gibberellin A_{12}-aldehyde in *Gibberella fujikuroi* strain REC-193A. *Phytochemistry* **12**, 2173–2179.

Bearder, J. R., MacMillan, J., Wels, C. M., Chaffey, M. B. and Pinney, B. O. (1974). Position of the metabolic block for gibberellin biosynthesis in mutant B1-41a of *Gibberella fujikuroi. Phytochemistry* **13**, 911–917.

Bearder, J. R., Dennis, F. G., MacMillan, J., Martin, G. C. and Phinney, B. O. (1975a). A new gibberellin (A_{45}) from seed of *Pyrus communis* L. *Tetrahedron Lett.*, 669–670.

Bearder, J. R., MacMillan, J. and Phinney, B. O. (1975b). Metabolic pathways from *ent*-kaurenoic acid to the fungal gibberellins in mutant B1-41a of *Gibberella fujikuroi. J. Chem. Soc. Perkin Trans.* I, 721–726.

Bearder, J. R., MacMillan, J., Wels, C. M. and Phinney, B. O. (1975c). The metabolism of steviol to 13-hydroxylated *ent*-gibberellanes and *ent*-kauranes. *Phytochemistry* **14**, 1741–1748.

Bearder, J. R., Frydman, V. M., Gaskin, P., MacMillan, J. and Phinney, B. O. (1976a). Fungal products. Part XVI. Conversion of isosteviol and steviol acetate into gibberellin analogues by mutant B1-41a of *Gibberella fujikuroi* and the preparation of [^{3}H] gibberellin A_{20}. *J. Chem. Soc. Perkin Trans.* I, 173–178.

Bearder, J. R., MacMillan, J. and Phinney, B. O. (1976b). Origin of oxygen atoms in the lactone bridge of C_{19}-gibberellins. *J. Chem. Soc. Chem. Commun.*, 834–835.

Bearder, J. R., MacMillan, J., von Cartenn-Lichterfelde, C. and Hanson, J. R. (1979). The removal of C(20) in gibberellins. *J. Chem. Soc. Perkin Trans.* I, 1918–1921.

Bell, R. A. and Turner, J. V. (1981). Stereoselective reduction of 3-keto gibberellin acids to 3β-ols using K-selectride with KH_2PO_4 buffer. *Tetrahedron Lett.*, 4871–4872.

Binks, R., MacMillan, J. and Pryce, R. J. (1969). Plant hormones VII. Combined gas chromatography–mass spectrometry of the methyl esters of gibberellins A_1 to A_{24} and their trimethylsilyl ethers. *Phytochemistry* **8**, 271–284.

Blechschmidt, S., Castel, U., Gaskin, P., Hedden, P., Graebe, J. E. and MacMillan, J. (1984). GC/MS analysis of the plant hormones in seeds of *Cucurbita maxima*. *Phytochemistry* **23**, 553–558.

Borrow, A., Jefferys, E. G., Kessell, R. H. J., Lloyd, E. C., Lloyd, P. B. and Nixon, I. S. (1961). The metabolism of *Gibberella fujikuroi* in stirred cultures. *Can. J. Microbiol.* **7**, 227–276.

Bowen, D. H., MacMillan, J. and Graebe, J. E. (1972). Determination of specific radioactivity of [^{14}C]-compounds by mass spectrometry. *Phytochemistry* **11**, 2253–2257.

Browning, G. and Saunders, P. F. (1977). Membrane localised gibberellins A_9 and A_4 in wheat chloroplasts. *Nature* **265**, 375–377.

Bukovac, M. J., Yuda, E., Murofushi, N. and Takahashi, N. (1979). Endogenous plant growth substances in developing fruit of *Prunus cerasus* L. VII. Isolation of gibberellin A_{32}. *Plant Physiol.* **63**, 129–132.

Cavell, B. D., MacMillan, J., Pryce, R. J. and Sheppard, A. C. (1967). Plant hormones V. Thin-layer and gas-liquid chromatography of the gibberellins; direct identification of the gibberellins in a crude plant extract by gas–liquid chromatography. *Phytochemistry* **6**, 867–874.

Coolbaugh, R. C. (1983). Early stages of gibberellin biosynthesis. *In*, "The Biochemistry and Physiology of Gibberellins" (A. Crozier, ed.) Vol I, pp. 53–98. Praeger, New York.

Coolbaugh, R. C., Heil, D. R. and West, C. A. (1982). Comparative effects of substituted pyrimidines on growth and gibberellin biosynthesis in *Gibberella fujikuroi*. *Plant Physiol.* **69**, 712–716.

Corcoran, M. R., Geissman, T. A. and Phinney, B. O. (1972). Tannins as gibberellin antagonists. *Plant Physiol.* **49**, 323–330.

Corey, E. J., Danheiser, R. L., Chandrasekaran, S., Keck, G. E., Gopalan, B. Larsen, S. D., Siret, P. and Gras, J.-L. (1978). Stereospecific total synthesis of gibberellic acid. *J. Am. Chem. Soc.* **100**, 8034–8036.

Cross, B. E., Galt, R. H. B. and Hanson, J. R. (1963). New metabolites of *Gibberella fujikuroi*. Part III. The structure of 7-hydroxykaurenolide. *J. Chem. Soc.*, 2944–2961.

Crozier, A. and Durley, R. C. (1983). Modern methods of analysis of gibberellins. *In* "The Biochemistry and Physiology of the Gibberellins" (A. Crozier, ed.) Vol. I, pp. 485–560. Praeger, New York.

Crozier, A., Aoki, H. and Pharis, R. P. (1969). Efficiency of counter current distribution, Sephadex G-10 and silicic acid partition chromatography in the purification and separation of gibberellin-like substances from plant extracts. *J. Exp. Bot.* **20**, 786–795.

Crozier, A., Zaerr, J. B. and Morris, R. O. (1980). High-performance steric exclusion chromatography of plant hormones. *J. Chromatogr.* **198**, 57–63.

Crozier, A., Zaerr, J. B. and Morris, R. O. (1982). Reversed-phase and normal-phase high performance liquid chromatography of gibberellin methoxycoumaryl esters. *J. Chromatogr.* **238**, 157–166.

Dockerill, B. and Hanson, J. R. (1978). The fate of C-20 in gibberellin biosynthesis. *Phytochemistry* **17**, 701–704.

Duri, Z. J., Fraga, B. M. and Hanson, J. R. (1981). Preparation of gibberellins A_9 and A_{20} from gibberellic acid. *J. Chem. Soc. Perkin Trans.* I, 161–164.

Durley, R. C. and Pharis, R. P. (1972). Partition coefficients of 27 gibberellins. *Phytochemistry* **11**, 317–326.

Durley, R. C. and Pharis, R. P. (1973). Interconversion of gibberellin A_4 to gibberellins A_1 and A_{34} by dwarf rice, cultivar Tan-ginbozu. *Planta* **109**, 357–361.

Durley, R. C., MacMillan, J. and Pryce, R. J. (1971). Investigation of gibberellins and other growth substances in seed of *Phaseolus multifloris* and *Phaseolus vulgaris* by gas chromatography and combined gas chromatography–mass spectrometry. *Phytochemistry* **11**, 317–326.

Durley, R. C., Crozier, A., Pharis, R. P. and McLaughlin, G. E. (1972). The chromatography of 33 gibberellins on a gradient eluted silica gel partition column. *Phytochemistry* **11**, 3029–3033.

Durley, R. C., Sassa, T. and Pharis, R. P. (1979). Metabolism of tritiated gibberellin A_{20} in immature seed of dwarf pea cv. Meteor. *Plant Physiol.* **64**, 214–219.

Eberle, J., Yamaguchi, I., Nakagawa, R., Takahashi, N. and Weiler, E. W. (1986). Monoclonal antibodies against GA_{13}-imide recognize the endogenous plant growth regulator, GA_4, and related gibberellins. *FEBS Letters* **202**, 27–31.

Evans, R., Hanson, J. R. and White, A. F. (1970). Studies in terpenoid biosynthesis. Part VI. The stereochemistry of some stages in tetracyclic diterpene biosynthesis. *J. Chem. Soc.* (C), 2601–2603.

Frydman, V. M., Gaskin, P. and MacMillan, J. (1974). Qualitative and quantitative analyses of gibberellins throughout seed maturation in *Pisum sativum* cv. Progress No. 9. *Planta* **118**, 123–132.

Fuchs, S. and Fuchs, Y. (1969). Immunological assay for plant hormones using specific antibodies to indole-acetic acid and gibberellic acid. *Biochim. Biophys. Acta* **192**, 528–531.

Fuchs, Y. and Gertman, E. (1974). Insoluble antibody columns for isolation and quantitative determination of gibberellins. *Plant Cell Physiol.* **15**, 629–633.

Fuchs, S., Haimovich, J. and Fuchs, Y. (1971). Immunological studies of plant hormones, detection and estimation by immunological assays. *Eur. J. Biochem.* **18**, 384–390.

Fujita, E. and Node, M. (1977). Synthesis of gibberellins. *Heterocycles* **7**, 709–752.

Fukui, H., Koshimizu, K., Usuda, S. and Yamazaki, Y. (1977a). Isolation of plant growth regulators from seeds of *Cucurbita pepo* L. *Agric. Biol. Chem.* **41**, 175–180.

Fukui, H., Nemori, R., Koshimizu, K. and Yamazaki, Y. (1977b). Structures of gibberellins A_{39}, A_{48} and A_{49} and a new kaurenolide in *Cucurbita pepo* L. *Agric. Biol. Chem.* **41**, 181–187.

Gaskin, P. and MacMillan, J. (1975). Polyoxygenated *ent*-kauranes and water-soluble conjugates in seed of *Phaseolus coccineus*. *Phytochemistry* **14**, 1575–1578.

Gaskin, P. and MacMillan, J. (1978). GC and GC–MS technique for gibberellins. *In* "Isolation of Plant Growth Substances" Society for Experimental Biology Seminar Series 4 (J. R. Hillman, ed.) pp. 79–95. Cambridge University Press, Cambridge.

Gaskin, P., MacMillan, J., Firn, R. D. and Pryce, R. J. (1971). "Parafilm": A convenient source of *n*-alkane standards for the determination of gas chromatographic retention indices. *Phytochemistry* **10**, 1155–1157.

Gaskin, P., Hutchison, M., Lewis, N., MacMillan, J. and Phinney, B. O. (1984). Microbiological conversion of 12-oxygenated and other derivatives of *ent*-kaur-16-en-19-oic acid by *Gibberella fujikuroi,* mutant B1-41a. *Phytochemistry* **23**, 559–564.

Gianfagna, T., Zeevaart, J. A. D. and Lusk, W. J. (1983). Synthesis of [^{2}H]gibberellins from steviol using the fungus *Gibberella fujikuroi*. *Phytochemistry* **22**, 427–430.

Glenn, J. L., Kuo, C. C., Durley, R. C. and Pharis, R. P. (1972). Use of insoluble polyvinylpyrrolidone for purification of plant extracts and chromatography of plant hormones. *Phytochemistry* **11**, 345–351.

Gräbner, R., Schneider, G. and Sembdner, G. (1976). Gibberelline XLIII Mitt. Fraktionierung von Gibberellinen, Gibberellinkonjugaten und anderen Phytohormonen durch DEAE-Sephadex-Chromatographie. *J. Chromatogr.* **121**, 110–115.

Graebe, J. E. (1982). Gibberellin biosynthesis in cell-free systems from higher plants. *In* "Plant Growth Substances 1982" (P. F. Wareing, ed.) pp. 71–80. Academic Press, London.

Graebe, J. E. and Ropers, H.-J. (1978). Gibberellins. In "Phytohormones and Related Compounds: A Comprehensive Treatise" (D. S. Letham, P. B. Goodwin and T. J. V. Higgins, eds) Vol. I, pp. 107–203. Elsevier/North-Holland, Amsterdam.

Graebe, J. E., Bowen, D. H. and MacMillan, J. (1972). The conversion of mevalonic acid into gibberellin A_{12}-aldehyde in a cell-free system from *Cucurbita pepo*. *Planta* **102**, 261–271.

Graebe, J. E., Hedden, P., Gaskin, P. and MacMillan, J. (1974). Biosynthesis of gibberellins A_{12}, A_{15}, A_{24}, A_{36} and A_{37} by a cell-free system from *Curcurbita maxima*. *Phytochemistry* **13**, 1433–1440.

Grob, K. and Grob Jr., K. (1974). Isothermal analysis on capillary columns without stream splitting. The role of the solvent. *J. Chromatogr.* **94**, 53–64.

Grove, J. F. (1961). The gibberellins. *Quart. Rev. Chem. Soc.* **15**, 56–70.

Grove, J. F., MacMillan, J., Mulholland, T. P. C. and Turner, W. B. (1960). Gibberellic acid. Part XVII. The stereochemistry of gibberic and epigibberic acid. *J. Chem. Soc.*, 3049–3057.

Hanson, J. R. (1965). Gibberellic acid. Part XXXI. The nuclear magnetic resonance spectra of some gibberellin derivatives. *J. Chem. Soc.*, 5036–5040.

Hanson, J. R. and Hawker, J. (1973). Preparation of [^{14}C]-gibberellic acid. *Phytochemistry* **12**, 1973–1975.

Hedden, P. (1983). *In vitro* metabolism of gibberellins. *In* "The Biochemistry and Physiology of gibberellins" (A Crozier, ed.) Vol. I, pp. 99–149. Praeger, New York.

Hedden, P. (1986). The use of combined gas chromatography–mass spectrometry in the analysis of plant growth substances. *In* "Gas Chromatography/Mass Spectrometry. Modern Methods of Plant Analysis, New Series Volume 3" (H. F. Linskens and J. F. Jackson, eds) pp. 1–22. Springer-Verlag, Berlin.

Hedden, P. and Graebe, J. E. (1981). Kaurenolide biosynthesis in a cell-free system from *Cucurbita maxima* seeds. *Phytochemistry* **20**, 1011–1015.

Hedden, P. and Graebe, J. E. (1982). The cofactor requirements for the soluble oxidases in the metabolism of the C_{20}-gibberellins. *J. Plant Growth Regulation* **1**, 105–116.

Hedden, P., MacMillan, J. and Phinney, B. O. (1974). Fungal products. Part XII. Gibberellin A_{14}-aldehyde, an intermediate in gibberellin biosynthesis in *Gibberella fujikuroi*. *J. Chem. Soc. Perkin Trans.* I, 587–592.

Hedden, P., MacMillan, J. and Phinney, B. O. (1978). The metabolism of the gibberellins. *Annu. Rev. Plant Physiol.* **29**, 149–192.

Hedden, P., Phinney, B. O., Heupel, R., Fujii, D., Cohen, H., Gaskin, P., MacMillan, J. and Graebe, J. E. (1982). Hormones of young tassels of *Zea mays*. *Phytochemistry* **21**, 391–393.

Hedden, P., Graebe, J. E., Beale, M. H., Gaskin, P. and MacMillan, J. (1984). The biosynthesis of 12α-hydroxylated gibberellins in a cell-free system from *Cucurbita maxima* endosperm. *Phytochemistry* **23**, 569–574.

Heftmann, E. and Saunders, G. A. (1978). Argentation thin-layer chromatography of the *p*-nitrobenzyl esters of gibberellins and their precursors. *J. Liquid Chromatogr.* **1**, 333–341.

Heftmann, E., Saunders, G. A. and Haddon, W. F. (1978). Argentation high-pressure liquid chromatography and mass spectrometry of gibberellin esters. *J. Chromatogr.* **156**, 71–77.

Hemphill, D. D., Jr., Baker, L. R. and Sell, H. M. (1973). Isolation of novel conjugated gibberellins from *Cucumis sativus* seeds. *Can. J. Biochem.* **51**, 1647–1653.

Hiraga, K., Yokota, T., Murofushi, N. and Takahashi, N. (1974). Isolation and characterization of gibberellins in mature seeds of *Phaseolus vulgaris*. *Agric. Biol. Chem.* **38**, 2511–2520.

Horgan, R. (1981). Modern methods for plant hormone analysis. *In* "Progress in Phytochemistry 7" (R. Rheinhold, J. B. Harborne and T. Swain, eds) pp. 137–170. Pergamon Press, Oxford.

Ingram, T. J. and Browning, G. (1979). Influence of photoperiod on seed development in the genetic line of peas *G2* and its relation to changes in endogenous gibberellins measured by combined gas chromatography–mass spectrometry. *Planta* **146**, 423–432.

Ishida, M., Suyama, K. and Adachi, S. (1980). Background contamination by phthalates commonly encountered in the chromatographic analysis of lipid samples. *J. Chromatogr.* **189**, 421–424.

Jensen, E., Crozier, A. and Monteiro, A. M. (1986). Analysis of gibberellins and gibberellin conjugates by ion-suppression reversed-phase high-performance liquid chromatography. *J. Chromatogr.* **367**, 377–384.

Jones, M. G. and Zeevaart, J. A. D. (1980). The effect of photoperiod on the levels of seven endogenous gibberellins in the long-day plant *Agrostemma githago* L. *Plant Physiol.* **69**, 660–662.

Jones, M. G., Metzger, J. D. and Zeevaart, J. A. D. (1980). Fractionation of gibberellins in plant extracts by reverse-phase high performance liquid chromatography. *Plant Physiol.* **65**, 218–221.

Jones, R. L. (1968). Aqueous extraction of gibberellins in peas. *Planta* **72**, 155–161.

Kagawa, T., Fukinbara, T. and Sumiki, Y. (1963). Thin layer chromatography of gibberellins. *Agric. Biol. Chem.* **27**, 598–599.

Kamiya, Y. and Graebe, J. E. (1983). The biosynthesis of all major pea gibberellins in a cell-free system from *Pisum sativum*. *Phytochemistry* **22**, 681–690.

Kende, H. (1967). Preparation of radioactive gibberellin A_1 and its metabolism in dwarf peas. *Plant Physiol.* **42**, 1612–1618.

Kirkwood, P. S. and MacMillan, J. (1982). Gibberellins A_{60}, A_{61} and A_{62}: Partial synthesis and natural occurrence. *J. Chem. Soc. Perkin. Trans.* I, 689–697.

Kirkwood, P. S., MacMillan, J. and Sinnot, M. L. (1980). Rearrangement of the lactone ring of gibberellin A_3 in aqueous alkali; participation of the ionised 3-hydroxy-group in an *anti* S_N2' reaction. *J. Chem. Soc. Perkin Trans.* I, 2117–2121.

Knox, J. P., Beale, M. H., Butcher, G. W. and MacMillan, J. (1987). Preparation and characterisation of monoclonal antibodies which recognise different gibberellin epitopes. *Planta* **170**, 86–91.

Koshioka, M., Harada, J., Takeno, K., Noma, M., Sassa, T., Ogiyama, K., Taylor, J. S., Rood, S. B., Legge, R. L. and Pharis, R. P. (1983a). Reverse-phase C_{18} high performance liquid chromatography of acidic and conjugated gibberellins. *J. Chromatogr.* **256**, 101–115.

Koshioka, M., Takeno, K., Beall, F. D. and Pharis, R. P.(1983b). Purification and separation of plant gibberellins from their precursors and glucosyl conjugates. *Plant Physiol.* **73**, 398–406.

Küllertz, G., Eckert, H. and Schilling, G. (1978). Quantitative gas chromatographic determination of gibberellic acid (GA_3) in nanogram quantities by electron capture detection. *Biochem. Physiol. Pflanz.* **173**, 186–187.

Kutschabsky, L. and Adam, G. (1983). Molecular and crystal structure of the phytohormone gibberellin A_3. *J. Chem. Soc. Perkin Trans.* I, 1653–1655.

Lewis, N. J. and MacMillan, J. (1980). Terpenoids. Part 8. Partial synthesis of *ent*-11β-, 12α- and 12β-hydroxykaur-16-en-19-oic acids from grandiflorenic acid. *J. Chem. Soc. Perkin Trans.* I, 1270–1278.

Lin, J.-T. and Heftmann, E. (1981). Adsorption and reversed-phase high performance liquid chromatography of gibberellins. *J. Chromatogr.* **213**, 507–510.

Lischewski, M. and Adam, G. (1980). Gibberellins LXII. Synthesis of 7-deoxygenated gibberellin A_3 compounds. *Tetrahedron* **36**, 1237–1244.

Lischewski, M., Adam, G., Liebisch, H.-W. and Pleiß, V. (1982). Partial synthesis of tritium labelled gibberellin-A_3. *J. Labelled Compd. Radiopharm.* **19**, 725–733.

Lombardo, L. (1982). Methylenation of carbonyl compounds with Zn-CH_2Br_2-$TiCl_4$. Applications to gibberellins. *Tetrahedron Lett.*, 4293–4296.

MacMillan, J. (1972). A system for the characterization of plant growth substances based upon direct coupling of a gas chromatograph, a mass spectrometer, and a small computer—recent examples of its application. *In* "Plant Growth Substances 1970" (D. J. Carr, ed.) pp. 790–797. Springer-Verlag, Berlin.

MacMillan, J. (1980). Partial synthesis of isotopically labelled gibberellins. *In* "Plant Growth Substances 1979" (F. Skoog, ed.) pp. 161–169. Springer-Verlag, Berlin.

MacMillan, J. (1984). Analysis of plant hormones and metabolism of gibberellins. *In* "The Biosynthesis and Metabolism of Plant Hormones" Society of Experimental Biology Seminar Series 23. (A. Crozier and J. R. Hillman, eds.) pp. 1–16. Cambridge University Press, Cambridge.

MacMillan, J. and Pryce, R. J. (1973). The gibberellins. *In* "Phytochemistry" (L. P. Miller, ed.) pp. 283–286. Van Nostrand-Reinhold, New York.

MacMillan, J. and Suter, P. J. (1963). Thin layer chromatography of gibberellins. *Nature* **197,** 790.

MacMillan, J. and Takahashi, N. (1968). Proposed procedure for the allocation of trivial names to the gibberellins. *Nature* **217**, 170–171.

MacMillan, J. and Wels, C. M. (1973). Partition chromatography of gibberellins and related diterpenes on columns of Sephadex LH-20. *J. Chromatogr.* **87**, 271–276.

Mandava, N. B. and Ito, Y. (1982). Separation of plant hormones by counter-current chromatography. *J. Chromatogr.* **247**, 315–325.

Mander, L. N. (1980). The total synthesis of gibberellins. *Search* **13**, 188–193.

Martin, G. C., Dennis, F. G., MacMillan, J. and Gaskin, P. (1977). Hormones in pear seeds I. Levels of gibberellins, abscisic acid, phaseic acid, dihydrophaseic acid, and two metabolites of dihydrophaseic acid in immature seeds of *Pyrus communis* L. *J. Am. Soc. Hort. Sci.* **102**, 16–19.

Melchers, F., Potter, M. and Warner, N. (eds) (1978). "Lymphocyte Hybridomas—Second Workshop on 'Functional Properties of Tumors of T and B Lymphocytes'," *Current Topics in Microbiology and Immunology,* **81**. Springer-Verlag, New York.

Metzger, J. D. and Zeevaart, J. A. D. (1980). Effect of photoperiod on the levels of endogenous gibberellins in spinach as measured by combined gas chromatography–selection ion current monitoring. *Plant Physiol.* **66**, 844–846.

Moler, G. F., Delongchamp, R. R., Korfmacher, W. A., Pearce, B. A. and Mitchum, R. K. (1983). Confidence limits in isotope dilution gas chromatography/mass spectrometry. *Anal. Chem.* **55**, 835–841.

Morris, R. O. and Zaerr, J. B. (1978). 4-Bromophenacyl esters of gibberellins, useful derivatives for high performance liquid chromatography. *Anal. Lett.* AII(i), 73–83.

Müller, P., Knöfel, H.-D., Liebisch, H. W., Miersch, O. and Sembdner, G. (1978). Untersuchungen zur Spaltung von Gibberellinglucosiden. *Biochem. Physiol. Pflanz.* **173**, 396–409.

Murofushi, N., Yamaguchi, I., Ishigooka, H. and Takahashi, N. (1976). Chemical conversion of gibberellin A_{13} to gibberellin A_4. *Agric. Biol. Chem.* **40**, 2471–2474.

Murofushi, N., Durley, R. C. and Pharis, R. P. (1977). Preparation of radioactive gibberellins A_{20}, A_5 and A_8. *Agric. Biol. Chem.* **41**, 1075–1079.
Murofushi, N., Nagura, S. and Takahashi, N. (1979). Metabolism of steviol by *Gibberella fujikuroi* in the presence of plant growth retardant. *Agric. Biol. Chem.* **43**, 1159–1161.
Murofushi, N., Sugimoto, H., Itoh, K. and Takahashi, N. (1980). A novel gibberellin, GA_{57}, produced by *Gibberella fujikuroi*. *Agric. Biol. Chem.* **44**, 1583–1587.
Murphy, P. J. and West, C. A. (1969). The role of mixed function oxidases in kaurene metabolism in *Echinocystis macrocarpa* Greene endosperm. *Arch. Biochem. Biophys.* **133**, 395–407.
Nadeau, R. and Rappaport, L. (1974). The synthesis of [^{3}H]gibberellin A_3 and [^{3}H]gibberellin A_1 by the palladium-catalyzed actions of carrier-free tritium on gibberellin A_3. *Phytochemistry* **13**, 1537–1545.
Nutbeam, A. R. and Briggs, D. E. (1982). Gibberellin–phenol interactions in plant extracts. *Phytochemistry* **21**, 2217–2223.
Phinney, B. O. (1984). GA_1, dwarfism and the control of shoot elongation in higher plants. *In* "The Biosynthesis and Metabolism of Plant Hormones" Society of Experimental Biology Seminar Series 23 (A. Crozier and J. R. Hillman, eds) pp. 17–41. Cambridge University Press, Cambridge.
Pitel, D. W. and Vining, L. C. (1970). Preparation of gibberellin A_1-3,4-^{3}H. *Can. J. Biol.* **48**, 259–263.
Pitel, D. W., Vining, L. C. and Arsenault, G. P. (1971). Improved methods for preparing pure gibberellins from cultures of *Gibberella fujikuroi*. Isolation by adsorption or partition chromatography on silicic acid and by partition chromatography on Sephadex columns. *Can. J. Biochem.* **49**, 185–193.
Potts, W. C. and Reid, J. B. (1983). Internode length in *Pisum*. III. The effect and interaction of the *Na/na* and *Le/le* gene differences on endogenous gibberellin-like substances. *Physiol. Plant.* **57**, 448–454.
Powell, L. E. and Tautvydas K. J. (1967). Chromatography of gibberellins on silica gel partition columns. *Nature* **213**, 185–193.
Pryce, R. J. (1973). Decomposition of aqueous solutions of gibberellic acid on autoclaving. *Phytochemistry* **12**, 507–514.
Rademacher, W. (1978). "Gaschromatographische Analyse der Veränderungen im Hormongehalt des wachsenden Weizenkorns." Dissertation, Georg-August-Universität, Göttingen, FRG.
Rademacher, W. and Graebe, J. E. (1979). Gibberellin A_4 produced by *Sphaceloma manihoticola*, the cause of the superelongation disease of cassava (*Manihot esculenta*). *Biochem. Biophys. Res. Commun.* **91**, 35–40.
Reeve, D. R. and Crozier, A. (1976). Purification of plant hormone extracts by gel permeation chromatography. *Phytochemistry* **15**, 791–798.
Reeve, D. R. and Crozier, A. (1977). Radioactivity monitor for high-performance liquid chromatography. *J. Chromatogr.* **137**, 271–282.
Reeve, D. R. and Crozier, A. (1978). The analysis of gibberellins by high performance liquid chromatography. *In* "Isolation of Plant Growth Substances" Society for Experimental Biology Seminar Series 4 (J. R. Hillman, ed.) pp. 41–77. Cambridge University Press, Cambridge.

Reeve, D. R. and Crozier, A. (1980). Quantitative analysis of Plant hormones. *In* "Hormonal Regulation of Development 1. Molecular Aspects of Plant Hormones. Encyclopedia of Plant Physiology New Series, Vol. 9" (J. MacMillan, ed.) pp. 203–280. Springer-Verlag, Berlin.

Reeve, D. R., Yokota, T., Nash, L. J. and Crozier, A. (1976). The development of a high performance liquid chromatograph with a sensitive on-stream radioactivity monitor for the analysis of ^{3}H and ^{14}C-labelled gibberellins. *J. Exp. Bot.* **21**, 1243–1258.

Rivier, L., Gaskin, P., Albone, K. S. and MacMillan, J. (1981). GC-MS identification of endogenous gibberellins and gibberellin conjugates as their permethylated derivatives. *Phytochemistry* **20**, 687–692.

Rosher, P. H., Jones, H. G. and Hedden, P. (1985). Validation of a radioimmunoassay for (+)-abscisic acid in extracts of apple and sweet pepper tissue using high-pressure liquid chromatography and combined gas chromatography–mass spectrometry. *Planta* **165**, 91–99.

Rowe, J. W. (ed.) (1968). "The Common and Systematic Nomenclature of Cyclic Diterpenes" Proposal IUPAC Comm. Org. Nomencl., 3rd rev. Forest Products Lab. USDA, Madison.

Sandberg, G., Dunberg, A. and Odén, P.-C. (1981). Chromatography of acid phytohormones on columns of Sephadex LH-20 and insoluble poly-*N*-vinylpyrrolidone, and application to the analysis of conifer extracts. *Physiol. Plant.* **53**, 219–224.

Schreiber, K., Weiland, J. and Sembdner, G. (1969). Isolierung von Gibberellin-A_8-*O*(3)-*β*-D-glucopyranoside aus Früchten von *Phaseolus coccineus. Phytochemistry* **9**, 189–198.

Schlenk, H. and Gellerman, J. L. (1960). Esterification of fatty acids with diazomethane on a small scale. *Anal. Chem.* **32**, 1433–1440.

Schneider, G. (1980). Gibberelline-80. Über strukturelle Einflüsse bei der Glucosylierung von Gibberellinen. *Tetrahedron* **37**, 345–349.

Schneider, G. (1983). Gibberellin conjugates. *In* "The Biochemistry and Physiology of Gibberellins" (A. Crozier, ed.) Vol. I, pp. 389–456. Praeger, New York.

Schneider, G. and Schliemann, W. (1979). Untersuchungen zur enzymatischen Hydrolyse von Gibberellin-*O*-glucosiden. II. Hydrolysegeschwindigkeiten von Gibberellin-2-*O*- und Gibberellin-3-*O*-glucosiden. *Biochem. Physiol. Pflanz.* **174**, 746–751.

Schneider, G., Jänicke, S. and Sembdner, G. (1975). Gibberelline XXXIV Mitt. Beitrag zur Gaschromatographie von Gibberellinen und Gibberellin-*O*-glucosiden – *N,O,*-Bis(trimethylsilyl)acetamid als Silysierungsreagens. *J. Chromatogr.* **109**, 409–412.

Schneider, G., Sembdner, G. and Schreiber, K. (1977). Gibberelline-L. Synthese von *O*(3)- und *O*(13)-glycosylierten Gibberellinen. *Tetrahedron* **33**, 1391–1397.

Schneider, G., Sembdner, G. and Phinney, B. O. (1984). Synthesis of GA_{20} glucosyl derivatives and the biological activity of some gibberellin conjugates. *J. Plant Growth Regul.* **3**, 207–215.

Seeley, S. D. and Powell, L. E. (1974). Gas chromatography and detection of microquantities of gibberellins and indoleacetic acid as their fluorinated derivatives. *Anal. Biochem.* **58**, 39–46.

Sembdner, G., Gross, G. and Schreiber, K. (1962). Die Dünnschichtchromagraphie von Gibberellinen. *Experientia* **18**, 584.

Sponsel, V. M. (1983). *In vivo* gibberellin metabolism in higher plants. *In* "The Biochemistry and Physiology of Gibberellins" (A. Crozier, ed.) Vol. I, pp. 151–250. Praeger, New York.

Sponsel, V. M. and MacMillan, J. (1977). Further studies on the metabolism of gibberellins (GAs) A_9, A_{20} and A_{29} in immature seeds of *Pisum sativum* cv. Progress No. 9. *Planta* **135**, 129–136.

Sponsel, V. M. and MacMillan, J. (1978). Metabolism of gibberellin A_{29} in seeds of *Pisum sativum* cv. Progress No. 9; use of $[^2H]$ and $[^3H]$ GAs, and the identification of a new GA catabolite. *Planta* **144**, 69–78.

Sponsel, V. M. and MacMillan, J. (1980). Metabolism of $[^{13}C_1]$ gibberellin A_{29} to $[^{13}C_1]$ gibberellin-catabolite in maturing seeds of *Pisum sativum* cv. Progress No. 9. *Planta* **150**, 46–52.

Turnbull, G. N., Crozier, A. and Schneider, G. (1986). HPLC-based methods for the identification of gibberellin conjugates: Metabolism of $[^3]$gibberellin A_4 in seedlings of *Phaseolus coccineus. Phytochemistry* **25**, 1823–1828.

Van den Dool, H. and Kratz, P. D. (1963). A generalization of the retention index system including linear temperature programmed gas-liquid partition chromatography. *J. Chromatogr.* **11**, 463–471.

Vining, L. C. (1971). Separation of gibberellin A_1 and dihydrogibberellin A_1 by argentation chromatography on a Sephadex column. *J. Chromatogr.* **60**, 141–143.

Wada, K., Imai, T. and Suibata, K. (1979). Microbial productions of unnatural gibberellins from (−)-kaurene derivatives in *Gibberella fujikuroi. Agric. Biol. Chem.* **43**, 1157–1158.

Weiler, E. W. and Wieczorek, U. (1981). Determination of femtomol quantities of gibberellic acid by radioimmunoassay. *Planta* **152**, 159–167.

Wels, C. M. (1977). High-sensitivity method of radio gas chromatography for ^3H- and ^{14}C-labelled compounds. *J. Chromatogr.* **142**, 459–467.

West, C. A. (1973). Biosynthesis of gibberellins. *In* "Biosynthesis and Its Control in Plants" (B. V. Milborrow, ed.) pp. 473–482. Academic Press, New York and London.

Yamaguchi, I., Takahashi, N. and Fujita, K. (1975a). Application of ^{13}C nuclear magnetic resonance to the study of gibberellins. *J. Chem. Soc. Perkin Trans.* I, 992–996.

Yamaguchi, I., Miyamoto, M. Yamane, H., Murofushi, N., Takahashi, N. and Fujita, K. (1975b). Elucidation of the structure of gibberellin A_{40} from *Gibberella fujikuroi. J. Chem. Soc. Perkin Trans.* I, 996–999.

Yamaguchi, I., Yokota, T., Yoshida, S. and Takahashi, N. (1979). High pressure liquid chromatography of conjugated gibberellins. *Phytochemistry* **18**, 1699–1702.

Yamaguchi, I., Kobayashi, M. and Takahashi, N. (1980). Isolation and characterization of glucosyl esters of gibberellin A_5 and A_{44} from immature seeds of *Pharbitis purpurea. Agric. Biol. Chem.* **44**, 1975–1977.

Yamaguchi I., Fujisawa, S. and Takahashi, N. (1982). Qualitative and semi-quantitative analysis of gibberellins. *Phytochemistry* **21**, 2049–2055.
Yokota, T. and Takahashi, N. (1981). Gibberellin A_{59}: A new gibberellin from *Canavalia gladiata. Agric. Biol. Chem.* **45**, 1251–1254.
Yokota, T., Murofushi, N., Takahashi, N. and Tamura, S. (1971). Gibberellins in immature seeds of *Pharbitis nil.* Part III. Isolation and structures of gibberellin glucosides. *Agric. Biol. Chem.* **35**, 583–595.
Yokota, T., Yamazaki, S., Takahashi, N. and Itaka, Y. (1974). Structure of pharbitic acid, a gibberellin-related diterpenoid. *Tetrahedron Lett.* 2957–2960.
Yokota, T., Hiraga, K., Yamane, H. and Takahashi, N. (1975). Mass spectrometry of trimethylsilyl derivatives of gibberellin glucosides and glucosyl esters. *Phytochemistry* **14**, 1569–1574.
Yokota, T., Reeve, D. R. and Crozier, A. (1976). The synthesis of (^{3}H)-gibberellin A_9 with high specific activity. *Agric. Biol. Chem.* **40**, 2091–2094.
Yokota, T., Murofushi, N. and Takahashi, N. (1980). Extraction, purification, and identification. *In* "Hormonal Regulation of Development 1. Molecular Aspects of Plant Hormones." Encyclopedia of Plant Physiology New Series, Vol. 9. (J. MacMillan, ed.) pp. 113–201. Springer-Verlag, Berlin.
Zavala, M. E. and Brandon, D. L. (1983). Localization of a phytohormone using immunocytochemistry. *J. Cell Biol.* **97**, 1235–1239.
Ziegler, R. S., Powell, L. E. and Thurston, H. D. (1980). Gibberellin A_4 production by *Sphaceloma manihoticola*, causal agent of cassava superelongation disease. *Phytopathology* **70**, 589–593.

3

Abscisic Acid and Related Compounds

Steven J. Neill and Roger Horgan

I. INTRODUCTION

Abscisic acid (ABA, **1**) is the best characterized inhibitory compound to have been isolated from higher plants. A recent book edited by Addicott (1983) describes the current state of knowledge regarding the physiology and biochemistry of ABA.

Although the full physiological role of ABA in plant growth and development remains to be elucidated, there is now little doubt that ABA plays an important role in plant–water relations. The observations of Hiron and Wright (1973) that a period of water stress caused a large increase in the ABA content of leaves heralded a period of intense study into the relationship between ABA production and water stress. It is of interest to note that this discovery coincided with the early development of physical methods for measuring ABA in plants. Thus, it is probably true to say that there are more reliable quantitative data for ABA in plants than for any of the other endogenous plant growth regulators. Nevertheless there are numerous examples in the literature of attempts to measure ABA by physical methods which fall outrageously short of any accepted analytical standards. This is particularly surprising since good analytical methods for ABA have been available for a number of years: see, for example, the reviews by Saunders (1978) and Dörffling and Tietz (1983).

The purpose of this chapter is to provide the reader with a detailed account of the physical methods currently available for ABA analysis and to attempt to define what in the authors' opinion are acceptable analytical standards for ABA measurements. The appendix contains full experimental details of two methods of ABA analysis, which in the experience of the

The Principles and Practice of Plant Hormone Analysis
0-12-198375-7

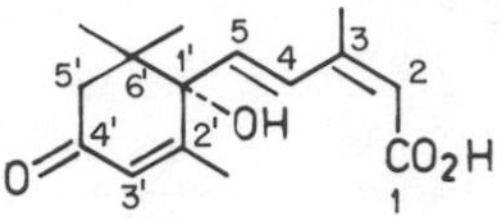

1 ABA 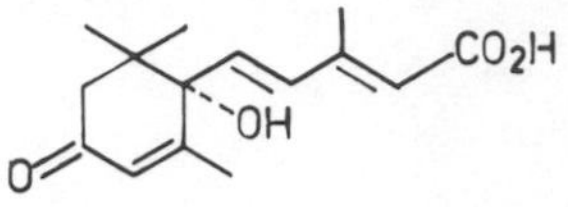 **2** *t*-ABA

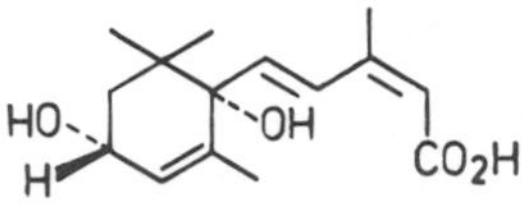

3 *GS*-diol 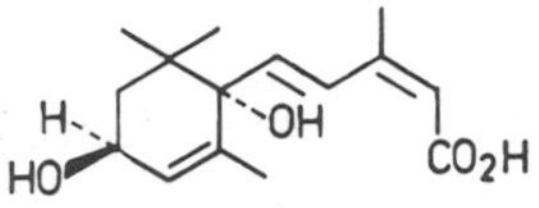 **4** *trans*-diol

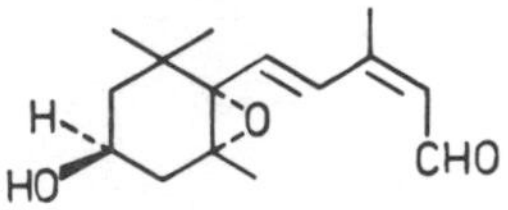

5 Xanthoxin 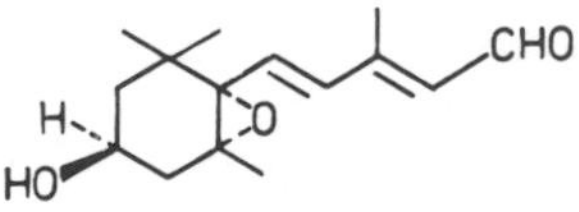 **6** *t*-xanthoxin

authors and numerous other workers have been found to be precise and accurate. We do not intend to discuss the use of bioassays in this chapter. Although bioassays may be very sensitive they are not selective for ABA and it is not possible to account readily for purification losses. However, bioassays are the only means of detecting new compounds with growth-inhibitory or anti-transpirant activity. Interested readers are referred to Milborrow (1978) and Dörffling and Tietz (1983). Since several of the methods discussed in this chapter are applicable to the study of ABA biosynthesis and metabolism these aspects will be covered in the relevant sections.

Studies of ABA biosynthesis pose particularly difficult analytical problems. As with all studies of biosynthesis, establishing radiochemical purity of the product is of paramount importance. In the case of trace components such as plant hormones this can only be done with considerable difficulty. Although chemical methods have been proposed for the radiochemical analysis of biosynthesized ABA (Milborrow and Noddle, 1970) it has not been possible to check the results of these by independent methods of proven analytical quality. The pathway of ABA biosynthesis in plants is still uncertain. Details of the current knowledge of ABA biosynthesis may be found in reviews by Milborrow (1983a), Horgan *et al.* (1983) and Neill *et al.* (1984). The isolation of xanthoxin (**5**) by Taylor and Burden (1970) suggested that ABA might be an apocarotenoid. Although there were several early studies on xanthoxin in plants (Firn *et al.*, 1972; Taylor and Burden, 1972; Zeevaart, 1974) including a demonstration that it is readily converted to ABA in tomato cuttings (Taylor and Burden, 1973) there has been very little recent work in this area. In view of its possible role as a precursor of ABA, some details of analytical methods for xanthoxin are included in this chapter.

Much more is known about the metabolism of ABA, and the established metabolic pathways of endogenous and exogenous ABA are shown in Fig. 1. The major pathway involves hydroxylation of one of the 6'-*gem* methyl groups to form 6'-hydroxymethyl ABA (6'-OHMeABA, **7**). This compound rearranges spontaneously to form phaseic acid (PA, **8**) which is subsequently reduced to dihydrophaseic acid (DPA; **9**) and epi-DPA (**10**) (Walton, 1980; Milborrow, 1983b). 6'-OHMeABA has only been isolated once but its formation has been inferred by the presence of the conjugate β-hydroxy-β-methylglutaryl hydroxy ABA (HMG-HOABA, **11**) (Hirai *et al.*, 1978).

The second pathway involves conjugation through the C-1 carboxyl group to form ABA-glucose ester (ABAGE, **12**). Although ABAGE has only

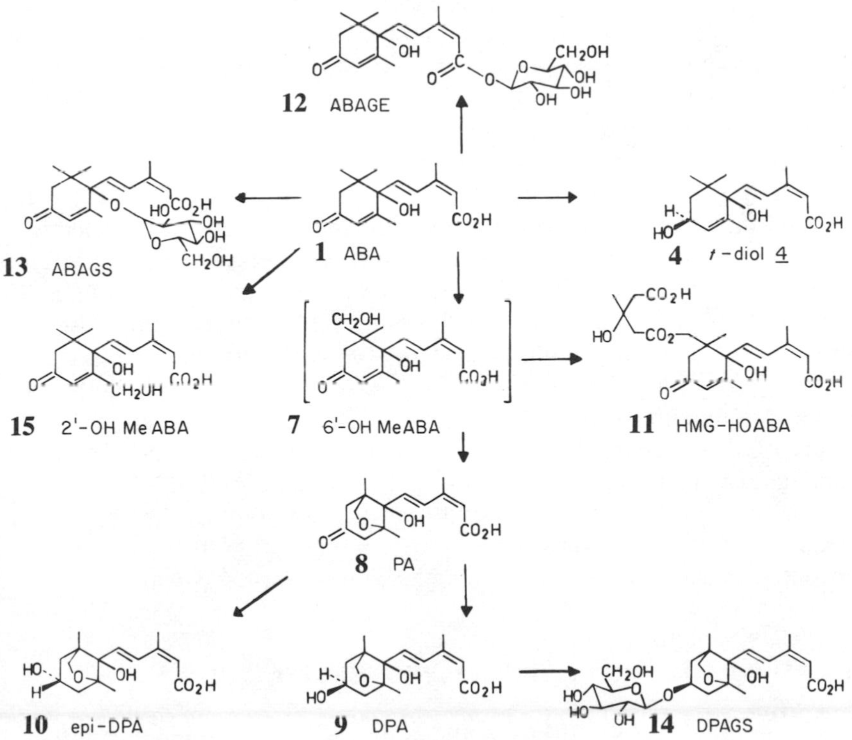

Fig. 1. ABA metabolism.

been identified in a few species (Koshimuzu *et al.*, 1968; Boyer and Zeevaart, 1982; Neill *et al.*, 1983) its presence in extracts has often been inferred by the presence of conjugated ABA, i.e. ABA released by alkaline hydrolysis. Exogenous ABA can also be conjugated with glucose to form the 1′-glucoside (ABAGS, **13**) (Loveys and Milborrow, 1981). The 4′-glucoside of DPA (DPAGS; **14**) has been found as both an exogenous and endogenous metabolite (Milborrow and Vaughan, 1982; Hirai and Koshimizu, 1983).

Milborrow (1983c) has reported that exogenous ABA can be reduced to the 1′,4′-*trans*-diol (*4*). The *cis*-diol (*3*) has been identified as a natural compound (Dathe and Sembdner, 1982). Lehmann *et al.* (1983) have reported the conversion of exogenous ABA to the unusual metabolite 2′-hydroxymethyl ABA (2′-OHMeABA; **15**) in suspension cultures of various species. In the great majority of metabolism studies racemic ABA has been applied to plant tissue. Consequently some of the metabolites observed may have been artefacts formed from (*R*)-ABA. Analytical methods for ABA metabolites are discussed in the appropriate sections.

The techniques that have had by far the greatest impact on the analysis of endogenous plant growth regulators in recent years have been combined gas chromatography–mass spectrometry (GC–MS), high-performance liquid chromatography (HPLC) and latterly immunoassays. Since the authors of this chapter have had considerable personal experience in the first two techniques, the emphasis of this chapter will be on these procedures.

Immunoassay is potentially the most powerful technique available for plant hormone analysis by virtue of its extremely high sensitivity, selectivity and relative technical simplicity. Many plant physiologists see this technique as revealing in the future "all they ever wanted to know about plant hormones". This optimism may well be justified, and since future studies into the inter-and intracellular localization of plant hormones will certainly require the superior sensitivity of immunoassays, the development and testing of immunoassay methods for ABA must be seen as an important aspect of ABA analysis. The pros and cons of the presently available immunoassay methods for ABA will be discussed in Section VIII.C. However, it should be noted at this point that the comments of the authors are based mainly on work published by others, since they have only limited first-hand experience of immunoassay techniques for ABA analysis.

II. PHYSICAL AND CHEMICAL PROPERTIES

The numbering system for ABA illustrated in (**1**) is the most commonly employed nomenclature although use is occasionally made of the carotenoid convention of numbering. ABA is generally considered to be a sesquiterpene, although what little evidence exists for its pathway of biosynthesis in higher plants is equally consistent with it being an apocarotenoid.

ABA possesses a single chiral centre at its 1′-carbon and the naturally occurring compound is exclusively the +(*S*)-enantiomer. Commercially available ABA is a racemic mixture, although stereospecific syntheses of both the (*R*)- and (*S*)-enantiomers have been reported (Kienzle *et al.*, 1978). The 2-*cis* double bond of ABA is isomerized by light to give the biologically inactive 2-*trans* isomer. Strictly speaking the name abscisic acid refers exclusively to the compound with a 2-*cis* double bond, although this is frequently referred to in the literature as *cis* or 2-*cis* ABA. In this chapter the term ABA will be used to refer to the 2-*cis* compound and its 2-*trans* isomer will be referred to as *t*-ABA. Commercially produced ABA is also marketed as mixed isomers, i.e. a mixture of ABA and *t*-ABA, at a somewhat lower price than the pure racemic ABA.

ABA is a C-15 carboxylic acid containing an unsaturated keto group and a tertiary hydroxyl group. The carboxyl group of ABA has a pK_a of 4.8. At pH values above about 5.8 it may be considered to be completely ionized and at pH values below 3.8 to exist almost exclusively in the undissociated form. Thus the solvent partitioning behaviour and chromatographic properties of ABA are strongly influenced by pH within this range.

From the analytical standpoint the important chemical reactions of ABA involve primarily the carboxyl group and to a lesser extent the keto group. The carboxyl group of ABA can be esterified by a variety of methods, the most useful of which is its conversion to the methyl ester by reaction with diazomethane. As will be discussed later the methyl ester of ABA (MeABA) (**16**) is an important derivative for the analysis of ABA by gas

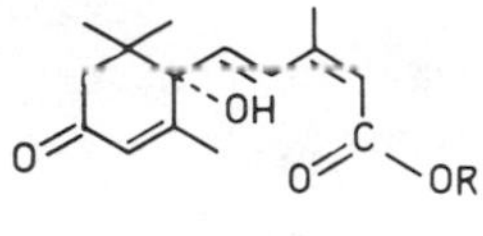

OH O O C OR

$R = CH_3$

16 Me ABA

chromatography (GC), GC–MS and HPLC. The keto group of ABA and MeABA can be reduced by sodium borohydride to give an approximately equimolar mixture of the epimeric 1′,4′-*cis* and 1′,4′-*trans* diols (**3**, **4**). When carried out on racemic ABA this reaction produces two pairs of diastereiomers. The diols can be readily oxidized back to ABA by active manganese dioxide in dry chloroform. The sequence of reactions methylation–reduction–oxidation, accompanied by thin-layer chromatography (TLC) analysis at each stage, has often been used to determine the radiochemical purity of ABA produced in biosynthetic studies (Milborrow and Noddle, 1970). Unfortunately there is no evidence from other analytical techniques against which this criteria of radiochemical purity can be assessed.

ABA exchanges six hydrogen atoms with water at high pH via the several possible enol forms of the α,β-unsaturated keto group. At pH values >10 this exchange is rapid and may be used to prepare hexadeuterated ABA for use as an internal standard. The hydrogens exchanged via the enol form of the ketone are reasonably stable at neutral pH, whereas the hydroxyl and carboxyl hydrogens re-exchange. The elegant elucidation of the structure of phaseic acid by Milborrow (1972) involved the use of hexadeuterated ABA prepared in this way. Hexadeuterated ABA is not an ideal internal standard for ABA determinations by GC–MS because careful regulation of the pH of extracts is required and the long-term stability of the deuterium atoms is unknown. It is more appropriate to use trideuterated ABA ([$^{2}H_3$]ABA) as the incorporated deuterium atoms are much more stable.

The reaction sequence for the preparation of [$^{2}H_3$]ABA is shown in Fig. 18 (Appendix B). This method, which was originally developed for the production of high specific activity [^{3}H]ABA by Walton *et al.* (1977), forms the basis for the commercial production of this material. Details of the synthesis of [$^{2}H_3$]ABA for use as an internal standard are given in Appendix B.

Trimethylsilylation of ABA under mild conditions (e.g. with hexamethyldisilizane (HMDS)/trimethylchlorosilane (TMCS) in pyridine) produces mainly the trimethylsilyl (TMS) ester. Due to its extreme sensitivity to water this derivative is of minimal value for ABA analysis when compared to MeABA. Trimethylsilylation of ABA with more powerful reagents (e.g. bistrimethylsilylacetamide) produces mixtures of 1′-0,4′-0 (via enolization) and the individual 1′-0 and 4′-0 TMS ether/esters (**17**). The difficulty of controlling the products arising from this reaction makes it of little value for the derivatization of ABA or the additional derivatization of MeABA.

III. CHEMICAL SYNTHESES

Several chemical syntheses of ABA have been described in the literature (Cornforth *et al.*, 1968; Findlay and MacKay, 1971; Roberts *et al.*, 1968), including a stereospecific synthesis of (*R*)- and (*S*)-ABA (Kienzle *et al.*, 1978). For the preparation of large quantities of material and the isotopic labelling of ABA the synthesis of Roberts *et al.* (1968) is the method of choice. We have used this method to produce ABA on a 10 g scale and to introduce [^{14}C]-, [^{3}H]- and [^{2}H]-labels into ABA.

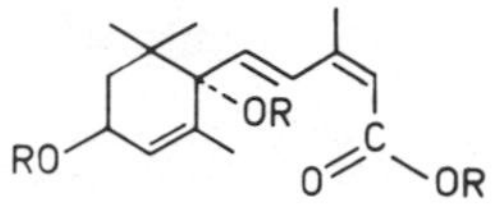

RO OR C O OR

R = $Si(CH_3)_3$

17 Tris TMS ABA

The key intermediate in the preparation of labelled ABA is the hydroxydiketone (**22**) shown in Fig. 18 in Appendix B. This can be prepared by following the detailed method given by Roberts *et al.* We have found that the chromatography step described in this paper can be omitted if the oily material from the oxidation is shaken with petrol ether until it becomes sufficiently solid to be filtered off. Recrystallization from toluene then yields pure hydroxydiketone. This material may then be used to prepare ^{3}H- or ^{2}H-labelled ABA as described in Appendix B, or for the preparation of ^{14}C-labelled ABA as described by Sondheimer and Tinelli (1971).

In the large-scale preparation of ABA we have found that effective separation of the mixture of ABA and *t*-ABA resulting from the Wittig reaction may be readily achieved by stirring the crude solid material with a suitable quantity of diethyl ether. In one large-scale preparation of ABA, 10 g of the crude mixture of isomers obtained from hydrolysis of the Wittig products was finely powdered and stirred with 500 ml of diethyl ether for 3 h at room temperature. Filtration followed by recrystallization from chloroform of the solid residue yielded 4.2 g of pure ABA. The solid residue resulting from evaporation of the ethereal filtrate was recrystallized from chloroform to yield 4.4 g of pure *t*-ABA.

IV. RESOLUTION OF ABSCISIC ACID ENANTIOMERS

The resolution of racemic ABA was first reported by Cornforth *et al.* (1967) using fractional crystallization of the (−) brucine salt. Sondheimer *et al.* (1971) have described a chromatographic method for the partial resolution of ABA and the hydroxy diketone (**22**) on acetyl cellulose columns. Complete resolution of the optically enriched fractions from these columns was achieved by selective solubilization. The resolved hydroxy diketone was used to prepare ^{14}C-labelled (*R*)- and (*S*)-ABA using the method of Roberts *et al.* (1968).

^{14}C-labelled (*R*)- and (*S*)-ABA have been prepared by Vaughan and Milborrow (1984) by resolution of *RS*-[2-^{14}C]MeABA 1′,4′-*cis*-diol (**3**) on an optically active HPLC column (Pirkle column) followed by oxidation and alkaline hydrolysis of the resolved diol esters. This would appear to be the most satisfactory technique for preparing small quantities of optically pure (*R*)- and (*S*)-ABA.

The use of an antibody column for resolving extremely small amounts of ABA has been described by Mertens *et al.* (1982). This technique is discussed further in Section VIII.C.

V. OPTICAL PROPERTIES

ABA shows a strong UV absorbance with a maximum at 260 nm (21 400) in acidic methanol and a shoulder at 240 nm (Milborrow, 1967). The ABA anion has a UV absorbance at 242.5 nm with an ε_{max} slightly higher than the free acid. The UV absorbance of MeABA is identical with that of the unionized acid.

The presence of the cyclohexenone chromophore in ABA causes the optically active (*R*)- and (*S*)-ABAs to exhibit exceptionally strong Cotton effects. The optical rotatory dispersion (ORD) curves show extremely large specific rotations at their extrema (+24 000 at 289 nm and −69 000 at 246 nm for (*S*)-ABA in acidified methanol). This property of ABA has been used as the basis of a method for its measurement (Milborrow, 1967). Due to lack of sensitivity in comparison with other techniques, it has been little used.

VI. MASS SPECTROMETRY

The electron impact (EI) mass spectrum of MeABA shows major ions at *m/z* 190, 162, 146, 134, 125, 112 and 91. The molecular ion at *m/z* 278 is very weak as is the ion at *m/z* 260, corresponding to the loss of water from the molecular ion. However, when using GC–MS the presence of the major ions, with correct relative intensities, at the GC retention time for MeABA may be taken as good evidence for its presence in a plant extract.

The fragmentation pattern of MeABA (**16**) has been studied in considerable detail by Gray *et al.* (1974) using metastable ions, isotopic labelling and high-resolution measurements. The fragmentation scheme is outlined in Fig. 2. The ions at *m/z* 222, 190, 162, 134, 106 and 91 occur on the same fragmentation pathway. Fragmentation via this pathway is initiated by cleavage of the 2′,3′-bond with elimination of isobutylene. Subsequent eliminations and cyclizations give rise to the other ions on this pathway, as shown in Fig. 2. The most abundant ion in the spectrum at *m/z* 190 has been shown to contain three of the four oxygen atoms originally present in the molecule. Knowledge of the oxygen atom content of the various ions on this pathway, as determined by Gray *et al.*, is of considerable value for investiga-

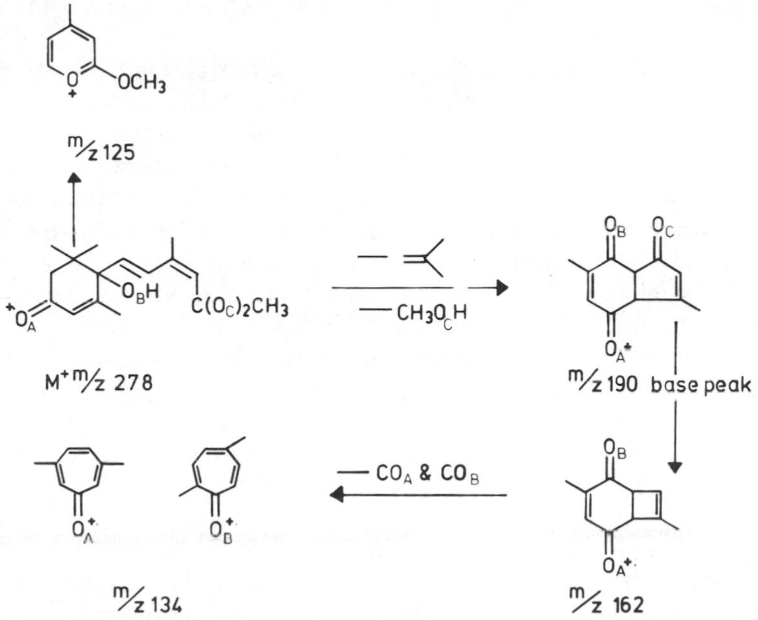

Fig. 2. Electron impact–mass spectral fragmentation pathways of MeABA.

tions into ABA biosynthesis using $^{18}O_2$ labelling. This technique has been utilized by Creelman and Zeevaart (1984) for an investigation into ABA and PA biosynthesis in water-stressed leaves of *Xanthium* and by Horgan *et al.* (unpublished) in studies of ABA biosynthesis in *Cercospora rosicola.* It should be noted that the major ions of this pathway also retain the side-chain 3-methyl group. Thus ABA labelled with deuterium in the side-chain methyl group is a valuable internal standard for the GC–MS determination of ABA. This topic will be covered in Section IX.

In addition to the major fragmentation pathway described above, the molecular ion of ABA fragments by an independent pathway involving cleavage of the 1′,5′-bond. This fragmentation gives rise to a cyclic ion at *m/z* 125 with retention of both ester oxygens as shown in Fig. 2. The retention of both of the carboxyl oxygen atoms of ABA in this ion makes it of value for ^{18}O studies.

The ion at *m/z* 125 in the spectrum of MeABA is also the major ion in the mass spectrum of the methyl esters of a number of compounds related to α-ionylidene acetic acid. Since these compounds may be conveniently labelled with deuterium in the side-chain methyl group, knowledge of their fragmentation pathways and that of ABA has enabled their biochemical conversion to ABA to be studied by mass spectrometry (Horgan *et al.*, 1983).

As will be discussed later, the accuracy of GC–single ion current monitoring (SIM) and GC–multiple ion monitoring (MIM) as techniques for measuring ABA in plant extracts depends on the uniqueness of the ion(s) chosen within the retention window of the compound under study. In this regard the base peak of MeABA at *m/z* 190 appears to be a fairly unique ion, at least within the normal GC retention window for MeABA. In the MS analysis of ABA from a wide variety of plant sources we have never observed significant interference with the *m/z* 190 ion by compounds co-eluting from the GC with MeABA.

Chemical ionization mass spectra (CI–MS) of MeABA and Me-*t*-ABA, using methane as the reagent gas, have been reported by Netting *et al.* (1982). As expected there is a considerable increase in the intensity of the pseudomolecular ion $[MH]^+$. However, whilst this is the base peak in the CI spectrum of Me-*t*-ABA, the base in the MeABA spectrum is at *m/z* 261, corresponding to loss of water from the protonated molecular ion. This ion has been used for the measurement of ABA levels in eucalyptus leaves by GC–CI–MS with $[^2H]$ABA as internal standard. In theory this technique should improve the selectivity of the MIM method by virtue of its use of an ion of higher mass than the *m/z* 190 ion used for the EI method.

It is of interest to note that the methane CI–MS of MeABA and Me-*t*-ABA show significant differences. This is in contrast to the EI–MS of these compounds which are sufficiently similar to necessitate the use of the combination of their GC retention times and mass spectra for full identification by conventional GC–MS.

By virtue of its inherent electron capturing properties MeABA might be expected to be a suitable compound for study by negative ion mass spectrometry. Rivier and Saugy (1986) investigated the negative ion CI–MS of MeABA using methane, isobutane and ammonia as reagent gases. In all cases the $[M]^-$ ion was the base peak. However, when ammonia was used as the reagent gas there was a dramatic increase in the ion current due to this ion. Using SIM it was possible to obtain a GC peak for the $[M]^-$ ion of ABA, with a signal/noise ratio of >3:1 for 0.3 pg of ABA injected onto a capillary column. The improvement in sensitivity and selectivity over the other methods of GC–MS determination of ABA was such that Rivier and Saugy were able to quantify ABA in a crude extract of maize roots by this technique, using a $[^2H_6]$ABA internal standard. It would thus appear that this is the most sensitive physical technique for the measurement of ABA.

The EI mass spectrum of MePA shows a molecular ion at *m/z* 294, which is somewhat stronger (relative intensity *ca.* 10%) than that of MeABA. Losses of H_2O, CH_3O and CH_3OH give rise to ions at *m/z* 276, 263, 262 and in combination *m/z* 244. The *m/z* 263 ion fragments to the *m/z* 139 ion with loss of the bridging CH_2O group. Loss of the side chain with retention of the tertiary oxygen atom gives rise to an ion at *m/z* 167 which further fragments with elimination of CO from the 4′-keto group to *m/z* 139. The major ion at *m/z* 125 derives in part from the side chain by a mechanism analogous to that for MeABA. However, deuterium labelling studies suggest that in the case of MePA this peak may contain ions derived both from the side chain and from the ring. The exact origin of the large ions at *m/z* 122 and 121 is unknown, although it has been shown that they both contain the side-chain methyl group.

The molecular ion of MeDPA is intermediate in intensity between that of MeABA and MePA. The fragmentation pathway of MeDPA is obviously very similar to that of MePA. The characteristic ion cluster at *m/z* 125, 122, 121 appears to have the same origin as that in MePA although in the case of MeDPA the intensity is shifted in favour of the 122, 121 pair.

The mass spectral fragmentation pattern of xanthoxin has not been elucidated in detail. Both xanthoxin (**5**) and its acetate (**18**) exhibit molecular ions with prominent fragments at M-CHO in both cases, with the additional loss of acetic acid in the case of the acetate. The prominent ion at *m/z* 149 (base peak in the acetate) found in both compounds is thought to include the complete side chain with the addition of three of the ring carbon atoms. An additional prominent ion at *m/z* 95 probably arises by side-chain cleavage leading to a cyclic ion similar to the *m/z* ion in the EI–MS of MeABA.

The EI mass spectrum of ABAGE-tetra-acetate (**19**) shows an extremely weak molecular ion at *m/z* 594. We have frequently failed to observe this ion and the correspondingly weak ion at *m/z* 534 due to the loss of acetic acid from the molecular ion. The most commonly occurring high-mass ion is found at *m/z* 441 and presumably arises via cleavage of the side-chain 5–1′ bond to form an ion analogous to that at *m/z* 125 in the EI spectrum of MeABA. Cleavage of the *O*–sugar bond gives rise to a series of ions at *m/z* 331, 271, 169 and 109. Ions arising from fragmentation of the aglycone occur at *m/z* 247, 190, 162, 134, 91 and are characteristic of the EI spectrum of ABA itself.

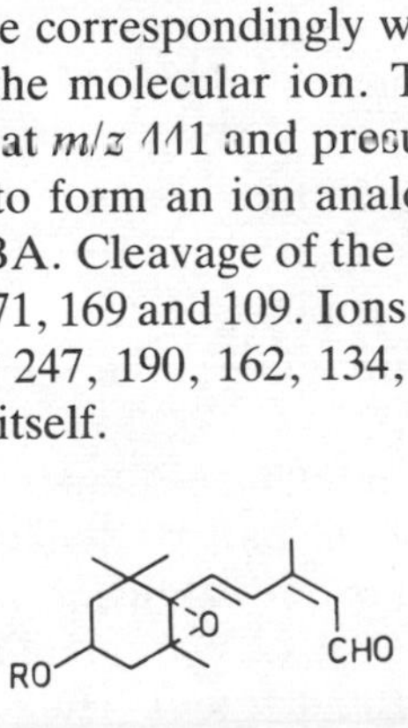

R = $COCH_3$
18 *O*-acetyl xanthoxin

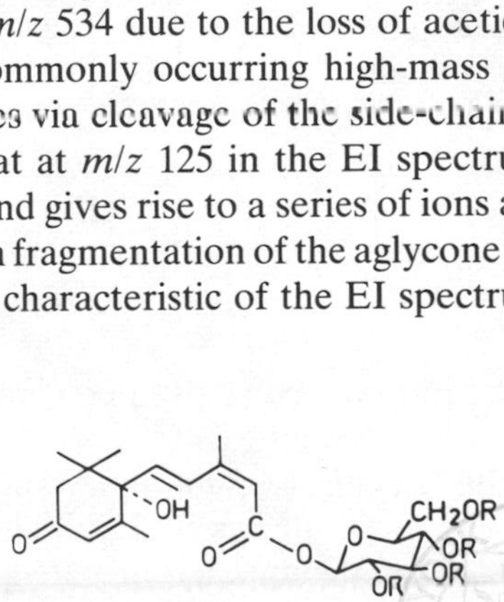

R = $COCH_3$
19 ABAGE tetra-acetate

The ammonia CI spectrum of ABAGE-tetra-acetate exhibits a strong pseudomolecular ion at *m/z* 612 $(M+NH_4)^+$. The major ions in this spectrum appear to arise by fragmentation of $(NH_4\text{-ions})^+$ involving the glucose moiety.

The mass spectra discussed above may be found in Appendix C, in a 10-peak format.

VII. ISOLATION

A. Extraction

If fresh material is not extracted immediately it is usually deep-frozen in liquid nitrogen and stored at −20°C. No data are available concerning possible hydrolysis of ABA-conjugates, further metabolism, etc. during cold storage.

As with other plant growth substances, various solvents have been used for the extraction of ABA and its metabolites from plant tissue. An ideal solvent would extract ABA in 100% yield without any rearrangement or breakdown of ABA or any of its metabolites, and with only minimal extraction of other plant constituents such as chlorophyll; 80% methanol has often been employed (Saunders, 1978) but Milborrow and Mallaby (1975) noted that in alkaline conditions "conjugated ABA" underwent transesterification to MeABA. We have confirmed that ABAGE does undergo transesterification and rearranges to several products in basic conditions (Neill *et al.*, 1983). We have used 80% acetone in 0.1M acetic acid and other workers have used acidic acetone for the extraction of ABA and its metabolites (e.g. Boyer and Zeevaart, 1982; Hirai and Koshimizu, 1983; Milborrow and Vaughan, 1982). Even though there is no way of assessing extraction efficiency, 80% acetone appears to be a suitable extraction solvent as re-extraction of tissue residues with further acetone did not improve yields of ABA or ABA plus labelled metabolites. Grinding the tissue with solvent in a mortar and pestle, Polytron or Waring blender followed by filtration is sufficient; longer extraction times do not increase yields.

Other solvents have also been employed as extractants. Ethanol (Most, 1971), 80% ethanol (Tietz *et al.*, 1979) and aqueous buffer (Rivier *et al.*, 1977) have been used for ABA and its metabolites. Mixtures of solvents have been used, such as boiling chloroform/methanol, (2:1, v/v) (Hillman *et al.*, 1974), methanol/ethyl acetate/acetic acid, (50:50:1, v/v) (Hubick and Reid, 1980) and methanol/ethyl acetate/formic acid, (50:50:1, v/v) (Loveys, 1977). Loveys found that for chloroplast extracts the latter solvent mixture yielded 50% more ABA than 80% methanol but gave no increased yield with leaf extracts. Methanol/chloroform/ammonium hydroxide (12:5:3, v/v) has been employed as an extraction solvent (Beardsell and Cohen, 1975); the high pH of this mixture may well lead to transesterification and hydrolysis of ABAGE. Walton *et al.* (1979) found water-saturated ethyl acetate a more convenient solvent than methanol and a three-phase extraction/partition system has been described (Liu and Tillberg, 1983).

The addition of antioxidants such as BHT (2,6-di-tert-butyl-4-methylphenol) at concentrations up to 100 mg l^{-1} (Zeevaart and Milborrow, 1976; Setter *et al.*, 1981) has been recommended. It is not clear how, or even whether or not, these compounds increase the yield of extracted ABA.

Plant extracts are usually reduced to aqueous solution or dryness prior to solvent partitioning or chromatography. The simple procedure of freezing and thawing followed by centrifugation to remove precipitated lipid material is usually very effective in reducing the weight of an extract with minimal loss of ABA. Saunders (1978) has recommended adjusting the extract pH to *ca.* 8 prior to freezing to avoid co-precipitation of ABA with lipid material; obviously this is not advisable if ABAGE is to be extracted.

B. Solvent partitioning

Partitioning between immiscible solvents is often employed as a purification step and can considerably reduce the dry weight of an extract. ABA is a weak acid having a pK_a of 4.8 and its distribution between water and various solvents thus depends upon the pH. At a pH below 3 ABA is almost completely undissociated and soluble in organic solvents such as ether; at a pH above 9 ABA is almost completely dissociated, being soluble in water but virtually insoluble in non-polar solvents such as ether. The solubility of ABA increases with increasing solvent polarity and this is reflected by the increasing partition coefficients in Table I. The dissociation of ABA can be exploited in a purification procedure. Extraction of an aqueous sample at pH 8 with ether removes basic and neutral ether-soluble compounds. If the pH of the aqueous residue is re-adjusted to pH 3 an acidic ether fraction can be prepared. This contains ABA and other organic ether-soluble acids, polar water-soluble compounds remaining in the aqueous phase. Fundamental considerations of solvent partitioning are discussed by Yokota *et al.* (1980).

The solubility of PA and DPA also depend on the pH. However, PA and especially DPA are more polar than ABA and thus less soluble in solvents such as ether, as can be seen from their partition coefficients in Table I. The solubility of ABAGE is not affected by pH because the carboxyl group is

Table 1. Partition coefficients ($K_a = C_{org}/C_{aq}$) of ABA and its metabolites between water and various solvents.

Solvent	ABA pH 2.5	ABA pH 9	PA pH 2.5	DPA pH 2.5	ABAGE pH 7
Hexane	0.01[b]	0.014[b]	—	—	—
Ether	3.3[b] 5[c] 2.3[e]	0.016[b] 0.007[c]	0.75[c]	0.11[e] 0.15[c]	0.03[a] 0.01[c]
Dichloromethane	0.95[b] 2.3[e]	0.008[b]	0.45[e]	0.12[e]	0.02[c]
Chloroform	3[b]	0.003[b]	—	—	—
Ethyl acetate	10[b] 10.1[e]	0.003[b]	4.3[e]	1.04[e]	0.135[a]
Water-saturated ethyl acetate	22[c]	0.14[c]	6.5[c]	0.9[c]	0.16[c]
Butanone	—	—	—	—	0.66[d]
1-Butanol	5.25[b]	0.46[b]	—	—	—
Water saturated 1-Butanol	52[b] 56[c]	0.47[b] 0.26[c]	34[c]	28[c]	3.1[d]

[a]Boyer and Zeevaart (1982), [b]Ciha *et al.* (1977), [c]Neill and Horgan, unpublished, [d]Neill *et al.* (1983), [e]Wealton *et al.* (1979).

esterified. ABAGE is less soluble in organic solvents than DPA, presumably due to the presence of the glucose moiety. However, it should be noted that ABAGE is not totally insoluble in ethyl acetate, having a partition coefficient of 0.16; three partitions with ethyl acetate would thus remove 35% of the ABAGE. Consequently any figures for conjugated ABA determined after ethyl acetate partitioning will be considerable underestimates.

The solubility of the aldehyde xanthoxin is not affected by pH and its polarity is lower than that of ABA. Consequently its solubility in organic solvents is greater. Firn *et al.* (1972) partitioned an 80% methanol extract with light petroleum. The methanol layer was then adjusted to 50% and re-extracted with light petroleum. After reduction of the methanol layer and re-dissolution in 5% sodium sulphate, xanthoxin was extracted with ether. Zeevaart (1974) used a modification of this procedure and Böttger (1978) partitioned an aqueous sample five times with ethyl acetate to remove the xanthoxin.

Recently Liu and Tillberg (1983) have suggested a three-phase technique involving partition between acidic buffer, ether and alkaline buffer contained in a dialysis bag. After 3 h stirring the alkaline buffer is then re-extracted with ether after stirring. Dumbroff *et al.* (1983) have recommended the use of "Clin-Elut" tubes for solvent partitioning, eliminating the need for separating funnels and providing filtration of the extract through an inert matrix.

In the authors' experience, when extract sizes are small solvent partitioning can conveniently be performed in Teflon-capped glass centrifuge tubes. This eliminates the need for separating funnels, breaks emulsions, reduces the time taken and increases recovery.

C. Column chromatography

Numerous open-column purification procedures for ABA have been reported in the literature but the advance of HPLC technology has rendered many of them redundant. Those briefly discussed below may still prove to be of value, particularly for large-scale extractions.

Considerable dry-weight reductions can be afforded by charcoal–celite chromatography. However, recoveries are variable and usually low. Charcoal columns eluted with increasing concentrations of acetone in water have been used to purify ABA and its metabolites (Harrison and Walton, 1975) and ABA and xanthoxin (Zeevaart, 1974).

Adsorption chromatography using silica gel has been used to purify ABA (see Yokota *et al.*, 1980). Hsu (1979) used silicic acid columns eluted with chloroform/methanol. Usually adsorption chromatography is performed on thin layers (see Section VII.D).

Various forms of Sephadex have been used for ABA purification. Sweetser and Vatvars (1976) eluted ABA from Sephadex G-25 with 20% methanol in water adjusted to pH 3 with sulphuric acid. ABA eluted prior to IAA in this system. Sephadex G-10 eluted with 0.25 M phosphate buffer has also been used as a clean-up step (Durley *et al.*, 1982). ABA and PA were eluted first, followed by IAA. Coloured impurities remained on the column. Sephadex LH-20 has also been used. Steen and Eliasson (1969) separated ABA and IAA with 70% ethanol and ABA and its glucose-ester can be resolved by elution with 35% ethanol in 0.1 M acetic acid (Neill *et al.*, 1983). Andersson *et al.* (1978) used two LH-20 columns in series, the first being eluted with 0.05 M phosphate buffer, pH 3, the second with methanol. Smaller extracts were chromatographed on a combined celite insoluble poly-*N*-vinylpyrrolidone (PVP)-LH-20 column with 0.05 M phosphate buffer prior to ether extraction and GC–MS analysis. In addition to molecular sieving the separation mechanisms of these Sephadex materials involve reversed-phase partitioning (LH-20) and adsorption/ion-exchange (G-10, G-25).

PVP is perhaps the most widely used material for bulk purification (Lenton *et al.*, 1971; Glenn *et al.*, 1972). PVP strongly binds to phenolic compounds and substantial dry-weight reductions can be achieved with excellent recoveries. Phosphate buffer at pH 6 is usually used as the eluent (Saunders, 1978). 0.1 M acetic acid can be used to elute ABAGE from PVP and ABA and its metabolites are eluted with three column volumes of this solvent (Neill, unpublished). Methanol has been used as a solvent (Mousdale and Knee, 1979; Durley *et al.*, 1982) but the highest dry-weight reductions are found when ABA is eluted as its ammonium salt (Saunders, 1978; Durley *et al.*, 1982).

Pre-packed chromatography cartridges (Sep-Pak, Waters Associates) have been used for ABA purification and are likely to find increased use in the future. They possess good separatory efficiency and sample capacity, and sample preparation time is greatly reduced. Reversed-phase C_{18}-Sep-Pak cartridges, eluted with methanol and acetic or phosphoric acid mixtures, have been used to purify ABA from *Cercospora* culture filtrates (Norman *et al.*, 1982; Griffin and Walton, 1982). The concentration of ABA in these extracts was of course very high and the ABA was subsequently analysed by reversed-phase HPLC. ABA and its metabolites have been "cleaned-up" on C_{18}-Sep-Pak cartridges prior to further reversed-phase HPLC purification (Milborrow and Vaughan, 1982; Pierce and Raschke, 1981). Whether or not C_{18}-cartridges would be suitable for purification directly prior to GC analysis remains to be seen. Hubick and Reid (1980) have described the use of silica Sep-Paks eluted with a series of organic solvents as a rapid purification procedure. After elution from the Sep-Pak the ABA was extracted with ether and quantified by GC with an electron capture detector (ECD). However, the identity of the putative MeABA was not confirmed by GC–MS and until it has been demonstrated that the ABA peak contains no impurities this method must be viewed with caution. Dumbroff *et al.* (1983) have also used silica Sep-Paks but the ABA was further purified by reversed-phase HPLC prior to quantitative analysis.

D. Paper and thin-layer chromatography

Paper chromatography (PC) has not been used extensively in the purification of ABA. However, it was frequently used for the fractionation of plant extracts prior to bioassay and led to the discovery of the "inhibitor β complex" which contained ABA. Some R_f values for ABA in various solvent systems are included in Dörffling and Tietz (1983) and Saunders (1978).

TLC on the other hand has been used frequently for the purification of ABA, and until the advent of HPLC was the method of choice. The resolving power of TLC is greater than that of PC although the sample capacity is not as high. TLC has been used for the purification of ABA metabolites (Harrison and Walton, 1975) and to follow ABA metabolism (e.g. Tietz *et al.*, 1979).

Solvent systems commonly employed include toluene/ethyl acetate/acetic acid in varying proportions, chloroform/acetic acid (98:2, v/v) and chloroform/methanol/water (75:22:3, v/v). The first two systems separate ABA and its *trans*-isomer whereas the last one does not—an important point to remember if *t*-ABA is added as an internal standard (see Section IX).

Dumbroff *et al.* (1983) have recently recommended the use of C_8 reversed-phase TLC plates with 60% methanol as solvent. R_f values of ABA and its metabolites on silica gel plates are shown in Tables II and III. R_f values on TLC are generally not very reproducible and reference compounds should therefore be used whenever possible.

The plates, especially homemade ones, should be prewashed in methanol/acetic acid (98:2, v/v) before use to remove impurities. In our experience plates do not require activation. Plates coated with a 0.5 mm layer of silica gel containing a fluorescence indicator are generally used, the ABA being detected as a quenching spot under UV light. When extracts are being purified by TLC, ABA marker spots must be kept separate from the extract so as to avoid contamination. This can conveniently be achieved by scoring the TLC plates with a needle prior to sample application. Plates should be developed in a saturated atmosphere (easily obtained by lining the tanks with chromatography paper) so as to avoid irregular solvent migration. After development, bands opposite the ABA marker spots should be eluted in solvents such as water-saturated ethyl acetate or acetone/methanol (1:1, v/v). Removal of the silica gel can be achieved by centrifugation or more conveniently by filtration through glass wool packed into a Pasteur pipette. As pointed out by Saunders (1978), plates should be eluted when they are barely dry in order to improve elution efficiency.

After TLC most samples will be amenable to analysis by GC–ECD following methylation. If, however, a sample is still too "dirty" the MeABA can be further purified by TLC in a solvent system such as hexane/ethyl acetate (see Table III).

TLC has also been used to purify xanthoxin from plant extracts (Table IV). Usually hexane/ethyl acetate solvents in various proportions have been employed (Firn *et al.*, 1972; Zeevaart, 1974). Similar solvents have also been used for the *O*-acetyl and *O*-heptafluorobutyryl-derivatives of xanthoxin (**18**, **19**) prior to GC analysis (Taylor and Burden, 1972; Böttger, 1978).

Unless there are financial constraints, TLC is likely to be replaced wherever possible by HPLC. However, it may still be useful for assessing the value of various solvent systems for subsequent use in adsorption HPLC.

E. High-performance liquid chromatography

Since its introduction in the late 1960s, HPLC has found extensive use in all areas of biochemistry and is the most powerful chromatographic tool

Table II. TLC R_f values of ABA and its metabolites in various solvent systems.

Solvent	ABA	PA	DPA	ABAGE	*c*-Diol	*t*-Diol	DPA-GS	2′OH-Me-ABA	HMG-HOABA
Benzene/ethyl acetate/acetic acid (50:5:2, v/v)	0.2[j]	0.1[j]	0.02[j]						
Benzene acetic acid (5:2, v/v)	0.65[j]	0.45[j]	0.29[j]						
Toluene/ethyl acetate/acetic acid (50:30:4, v/v)	0.4[b] 0.5[a]	0.33[b]				0.19[g]	0.33[g]		
Toluene/ethyl acetate/acetic acid (50:30:4, v/v) × 2	0.7[i,k]	0.5[i,k]	0.15[i,k]						
Chloroform/methanol/water (75:22:3, v/v)	0.7[j,a,h]	0.6[j] 0.5[a]	0.46[k]	0.36[a] 0.49[h]					
Toluene/ethyl acetate/methanol/acetic acid (50:30:7:4, v/v)	0.68[e]	0.53[e]	0.3[e]	0.1[e]				0.43[e]	
1-Propanol/ethyl acetate/water (3:2:1, v/v)	0.87[h]			0.83[h] 0.9[d]					
Chloroform/methanol/water (67:30:3, v/v)							0.35[f]		
Hexane/ethyl acetate/acetic acid (7:12:1, v/v)									0.24[c]

[a]Boyer and Zeevaart (1982); [b]Dörffling and Tietz (1983); [c]Hirai *et al.* (1978); [d]Koshimizu *et al.* (1968); [e]Lehmann *et al.* (1983); [f]Milborrow and Vaughan (1982) [g]Neill (unpublished); [h]Neill *et al.* (1983); [i]Tietz *et al.* (1979); [j]Walton and Sondheimer (1972); [k]Zeevaart and Milborrow (1976).

Table III. TLC R_f values of methylated ABA and metabolites.

Solvent	MeABA	MePa	MeDPA	Me-*c*-diol	Me-*t*-diol
Ethyl acetate/hexane (1:1, v/v)	0.67[e]			0.3[a]	0.44[a]
Ethyl acetate/hexane (3:2, v/v)			0.16[b] *epi*: 0.22[b]		
Ethyl acetate/hexane (2:1, v/v)	0.86[d]	0.74[e]	0.24[d]		
Ethyl acetate/hexane (2:1, v/v) × 5			0.38[f] *epi*: 0.52[f]		
Chloroform/methanol (1:1, v/v)				0.3[c]	0.6[c]

[a]Cornforth *et al.* (1967), [b]Hirai and Koshimizu (1983), [c]Dathe and Sembdner (1982), [d]Tietz *et al.* (1979), [e]Zeevaart (1974), [f]Zeevaart and Milborrow (1976).

available for the purification of plant hormones such as ABA. Some of its advantages over conventional chromatography are (i) reduced analysis time; (ii) far greater resolving power; (iii) increased ease of sample recovery, which is virtually quantitative. As long as the solvent is volatile enough then the column eluate is simply collected and the solvent evaporated; (iv) non-volatile samples can be analysed. In addition to its use in the preparative mode, HPLC can be used for quantitative analysis in conjunction with a suitable detector. Various HPLC detectors are available, but only UV detectors have been used for ABA analysis; their use is discussed in Section VIII.A.

Modern HPLC packing materials used for ABA analysis consist of irregular or spherical, totally microporous (5–10 μm diameter) silica particles with a large surface area (200–500 $m^2\ g^{-1}$). These stationary phases provide more efficient columns than pellicular materials and have a much higher sample capacity.

Table IV. TLC R_f values of xanthoxin and derivatives.

Solvent	XAN	XAN-0-Ac	XAN-0-PFB
Chloroform/methanol/water (75:22:3, v/v)	0.83[d]		
Ethyl acetate/hexane (1:4, v/v)			*c* – 0.32[a] *t* – 0.37[a]
Ethyl acetate/hexane (2:3, v/v)	0.2[a]	0.66[d]	
Ethyl acetate/hexane (1:1, v/v)	0.2[c]	0.5[c]	

[a]Böttger (1978), [b]Dörffling and Tietz (1983), [c]Taylor and Burden (1972), [d]Zeevaart (1974).

A wide range of packing materials is available and the use of HPLC for analysis of ABA will be discussed for each particular mode of chromatography.

(1) Reversed-phase HPLC

In reversed-phase HPLC the stationary phase consists of microporous silica particles to which has been bonded a non-polar hydrocarbon phase. The mechanism of separation involves partition between the more polar mobile phase and the non-polar stationary phase. Thus the most polar compounds elute first, the least polar ones last. Reversed-phase materials with C_2, C_6, C_8, C_{18} and phenyl-bonded phases are available. The majority of ABA analyses have been performed with the C_{18} octadecylsilane (ODS) phases.

ODS-bonded materials are extremely useful for the purification of ABA from plant extracts. Usually the columns are eluted with gradients of methanol or ethanol and acidic water. ODS columns have high sample capacity and excellent resolving power for a wide range of compounds, especially when operated in the gradient elution mode. The relatively low polarity of most compounds present in an ether extract means that large injections in water can be made onto ODS columns (we normally inject 1 ml on to semipreparative columns). Due to its low aqueous solubility the extract tends to concentrate as a narrow band on top of the column and subsequent resolution is not impaired. As pointed out by Horgan (1981), up to 100 mg of a multicomponent ether extract can be fractionated on semipreparative (150 × 10 mm i.d.) columns. In our experience crude extracts of up to 5 g (fresh weight) of plant material can be chromatographed satisfactorily on these large columns. Smaller analytical columns (4.5 mm i.d.) are of limited value for purifying crude extracts although they are suitable for further purification of already semipurified extracts and for quantitative analysis.

The separation mechanism involved in reversed-phase chromatography is apparently not simply due to reverse partition but also involves interaction with residual adsorption sites on the silica support. This can lead to considerable batch-to-batch variation for the chromatography of cytokinins (see Chapter 5, Volume 2) but the effects appear to be minimal for ABA and its metabolites. Table V shows the retention times of ABA and its *trans*-isomer and the resolution of the two compounds on three ODS supports. The 150 × 4.5 mm i.d. columns were eluted with a linear gradient of 0–100% methanol in 0.1 M acetic acid over 30 min at 2 ml min^{-1}. There is little difference between the three materials although the retention times and resolution are greater on Zorbax-ODS. This is probably a consequence of the heavier loading of stationary phase on this support.

Table V. Resolution of ABA and *t*-ABA on reversed-phase HPLC supports.

HPLC support	Retention time ABA (min)	Retention time *t*-ABA (min)	Resolution
ODS-Spherisorb	10.9	10	2
ODS-Apex	11.5	9.9	2.1
ODS-Zorbax	12.65	11.1	3.5

150 × 4.5 mm i.d. columns were eluted with a linear gradient of 0–100% methanol in 0.1 M acetic acid over 30 min at 2 ml min^{-1}. Resolution (R_s) was calculated as $R_s = (R_t[\text{ABA}] - R_t[t\text{-ABA}])/W$, where R_t[ABA] and R_t[*t*-ABA] are the retention times of ABA and *t*-ABA respectively and W is the average peak width.

Some workers have used relatively large (35–70 μm) particles for semi-preparative work (e.g. Ciha *et al.*, 1977; Durley *et al.*, 1978; Zeevaart, 1983) as these packing materials are less expensive than microparticulate supports. However the sample capacity of microparticulate columns is often superior (Brenner, 1981) and in addition they offer higher efficiency.

There are many reports of ABA purification using 5–10 μm ODS-semipreparative and analytical columns eluted with acetic acid/ethanol (e.g. Pierce and Raschke, 1981; Setter *et al.*, 1981) or acetic acid/methanol gradients (e.g. Boyer and Zeevaart, 1982; Ciha *et al.*, 1977; Durley *et al.*, 1982; Neill *et al.*, 1983). A chromatogram showing the separation of ABA and its metabolites is shown in Fig. 3 while a typical trace of a plant extract is illustrated in Fig. 16 (Appendix A). 0.1 M acetic acid is typically used as the aqueous phase, although other buffers have been used. Wheaton and Bauscher (1977) used 0.015 M ammonium acetate, pH 5.1, as the aqueous phase. Phosphoric acid (Norman *et al.*, 1982) and formic acid (Taylor and Rossall, 1982) have also been used. In addition to gradient elution, ODS columns have also been run isocratically when chromatographing partially purified extracts (e.g. Liu and Tillberg, 1983; Bangerth, 1982) or in quantitative analysis (Griffin and Walton, 1983; Neill and Horgan, 1983; Norman *et al.*, 1982).

At low pH the ionization of ABA (and its acidic metabolites) is suppressed. At higher pH increasing proportions of the ABA will be dissociated and thus more polar. Consequently if ABA is chromatographed on an ODS support using methanol/water, pH 7, the elution time is greatly reduced. ABA and its *trans* isomer are nonetheless still resolved (Mousdale, 1981). By including either tetramethylammonium phosphate or tetrabutyl ammonium hydrogen sulphate as a counter-ion in the aqueous phase (Mousdale (1981) increased the retention time of ABA. Ion-pair chromatography may be of value in situations where ABA co-elutes during ion-suppression chromatography with UV absorbing or electron-capturing compounds.

Usually ABA is purified by reversed-phase HPLC as the free acid,

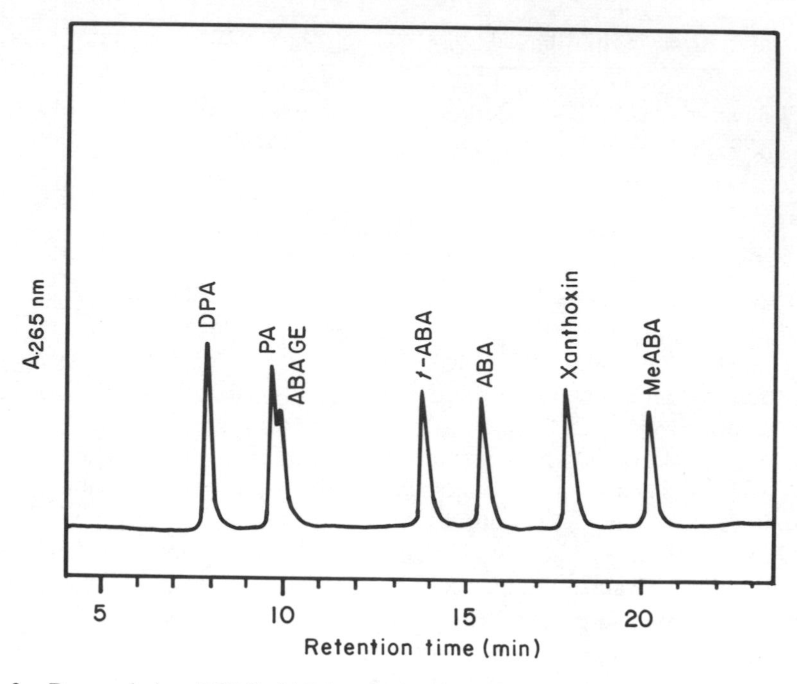

Fig. 3. Reversed-phase HPLC of ABA and related compounds. Column: 250 × 4.6 mm 5 μm ODS-Hypersil. Mobile phase: 30 min linear gradient of 0–100% methanol in 0.1 M aqueous acetic acid. Flow rate: 2 ml min^{-1}. Detector: absorbance monitor at 265 nm.

although extracts can be methylated prior to HPLC. Cargile *et al.* (1979) have described a purification procedure based on HPLC of methylated extracts on Bondapak C_{18} with 25% acetonitrile as eluent.

Reversed-phase HPLC is also very suitable for the purification of ABA metabolites. ABAGE has been purified from plant extracts using HPLC (Boyer and Zeevaart, 1982; Neill *et al.*, 1983) and ABAGS has been chromatographed on ODS using acetic acid/ethanol and acetic acid/acetonitrile mixtures (Loveys and Milborrow, 1981). PA and DPA have also been chromatographed on C_{18} reversed-phase materials using various solvents (Hirai and Koshimizu, 1983; Milborrow and Vaughan, 1982; Mapelli and Rocchi, 1983; Neill and Horgan, unpublished; Pierce and Raschke, 1981; Setter *et al.*, 1981; Zeevaart, 1983). The 4′-*O*-β-D-glucoside of DPA has recently been identified as an ABA metabolite and its HPLC purification described (Milborrow and Vaughan, 1982; Hirai and Koshimizu, 1983). Separation of ABA and some of its metabolites on an analytical ODS-Hypersil column is shown in Fig. 3. PA and ABAGE are not well resolved and hardly separate at all on a less efficient, semipreparative column. Poor resolution of PA and ABAGE has been observed by other authors (Setter *et al.*, 1981; Zeevaart, 1983) although Pierce and Raschke (1981) found that PA and an ABA conjugate probably representing ABAGE were resolved on an analytical C_{18} column eluted with an acetic acid/ethanol gradient. These differences may reflect differing selectivities of the reversed-phase materials.

There are no reports on HPLC of xanthoxin in the literature; its mobility on ODS-Hypersil is shown in Fig. 3.

(2) *Adsorption HPLC*

Adsorption HPLC using a silica support has been used both as a final purification step and to quantify ABA. The least polar compounds elute first, and there is considerable variation between different types of silica supports. Silica particles with the greatest surface area are the most active and adsorb compounds the most strongly.

Wheaton and Bauscher (1977) used chloroform/acetonitrile/acetic acid (15:15:0.4) to elute ABA and other workers have used similar solvents (Ciha *et al.*, 1977; Durley *et al.*, 1982). We routinely use water-saturated chloroform/acetic acid (100:2) as the eluting solvent. (see Fig. 4). Note that *t*-ABA elutes before ABA.

Less polar solvents such as hexane/2-propanol (95:5, v/v) (Neill *et al.*, 1983; Walton *et al.*, 1979) or acetonitrile/ethylene chloride (7:93, v/v) (Cargile *et al.*, 1979) are used to elute MeABA. With the hexane solvent the MeABA now elutes before Me-*t*-ABA (Fig. 5) whereas the order is reversed with the acetonitrile-based solvent.

Adsorption HPLC has also been used to chromatograph PA using an acidified acetonitrile/chloroform solvent (Durley *et al.*, 1982; Hirai and Koshimizu, 1983). We have used the same acidic chloroform solvent to analyse PA (Fig. 4), as described above for ABA. Stronger solvents are required to elute DPA and ABAGE from silica columns. These can conveniently be prepared by the addition of methanol to the chloroform/acetic acid solvent (see Figs 6 and 7). The methyl esters of PA and DPA can be chromatographed on silica using hexane/2-propanol mixtures (Fig. 8), although the percentage of 2-propanol is higher than that used to analyse MeABA and Me-*t*-ABA.

Xanthoxin can be analysed by HPLC using various combinations of the above solvents although the two isomers are not resolved. This separation can be achieved with a chloroform/acetonitrile (100:5, v/v) mobile phase (Fig. 9). After acetylation the *O*-acetyl derivative can be purified by silica adsorption HPLC using chloroform as the solvent; the two isomers do not separate.

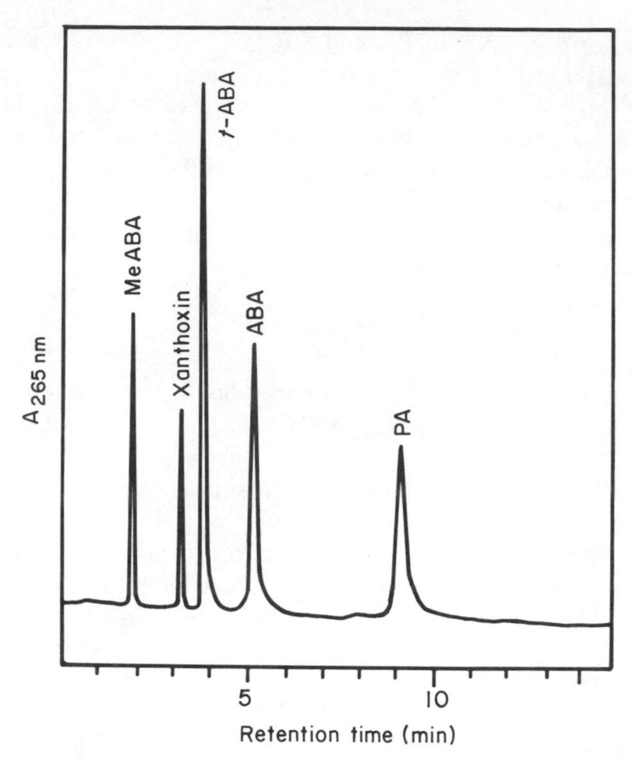

Fig. 4. Adsorption HPLC of ABA and related compounds. Column: 250 × 4.6 mm i.d. 5 μm Partisil. Mobile phase: chloroform/acetic acid (100:2, v/v). Flow rate: 3 ml min^{-1}. Detector: absorbance monitor at 265 nm.

(3) Normal-phase HPL

Normal-phase HPLC has also been employed. Ciha *et al.* (1977) used the amine bonded phase, μBondapak-NH_2, eluted with 45% acetonitrile in chloroform acidified with acetic acid. Mapelli and Rocchi (1983) have used a similar system to separate ABA, PA and DPA. ABA metabolites have also been purified on NH_2-columns using increasing concentrations of ethyl acetate in hexane as the eluent (Pierce and Raschke, 1981; Zeevaart, 1983).

PA and ABAGE, difficult to separate in reversed-phase, can be resolved by normal-phase HPLC. Setter *et al.* (1981) separated PA and an ABA conjugate (presumably ABAGE) by HPLC on NH_2-Bondapak using a solvent based on aqueous acetic acid, acetonitrile and ethanol.

In our laboratory we have used Partisil-PAC, a very useful polar bonded phase consisting of a mixture of cyano-propyl and amino groups. PA and ABAGE are very well resolved using a linear gradient of 0.1 M acetic acid in

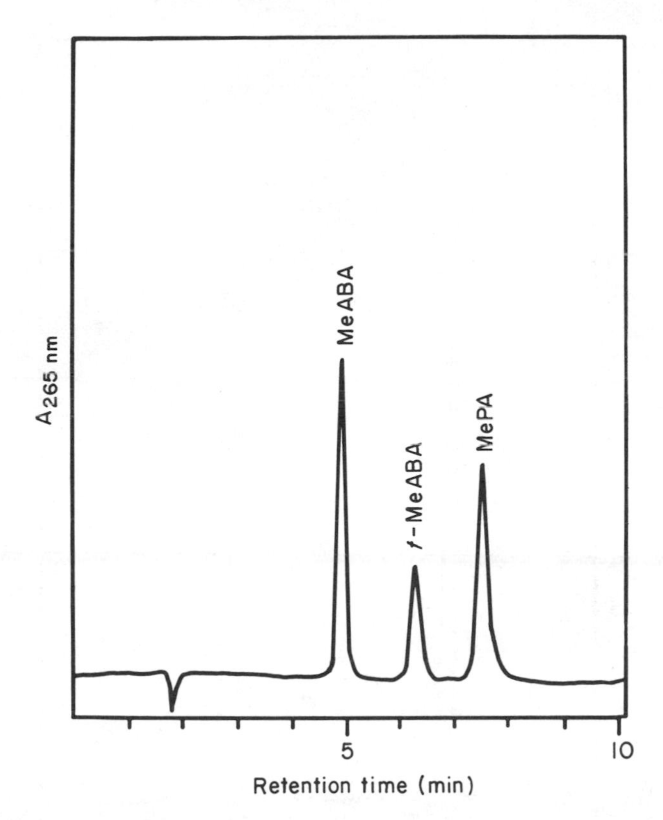

Fig. 5. Adsorption HPLC of MeABA and related compounds. Column: 250 × 4.6 mm i.d. 5 μm Spherisorb. Mobile phase: hexane/2-propanol (95:5, v/v). Flow rate: 3 ml min^{-1}. Detector: absorbance monitor at 265 nm.

acetonitrile. The separation of ABA and its metabolites is shown in Fig. 10. The separation mechanism probably involves ion exchange in addition to normal-phase partition. At low pH the NH_2 group will be protonated and exist in the NH_3^+ form, thus acting as an anion exchanger. PA, DPA and ABA still elute in order of increasing polarity. ABAGE is eluted first as it is not charged at any pH.

Xanthoxin cannot be chromatographed on Partisil-PAC and probably not on any $-NH_2$ bonded phases. One explanation for the failure of xanthoxin to elute may be that it condenses with the $-NH_2$ groups to form a Schiff's base.

PAC is also a useful stationary phase for the purification of MeABA, using hexane/ethanol (100:5, v/v) as the solvent. Velasco *et al.* (1978) have used PAC with a mobile phase of hexane/chloroform/methanol (75:20:5, v/v) to purify the *p*-nitrobenzyl ester of ABA.

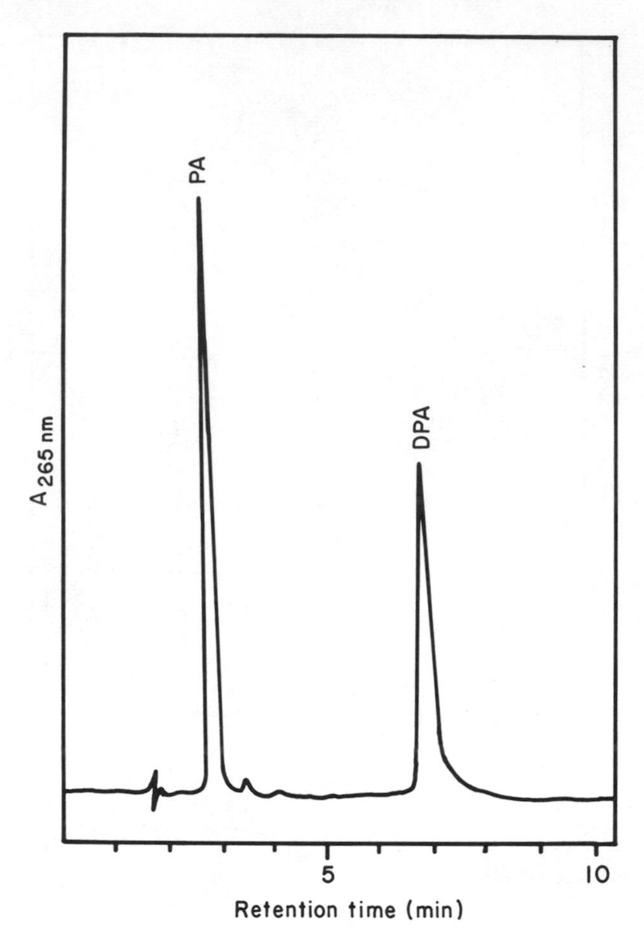

Fig. 6. Adsorption HPLC of ABA metabolites. Column: 250 × 4.6 mm i.d. 5 μm Partisil. Mobile phase: [chloroform/acetic acid (100:2, v/v)]/methanol (96:4, v/v). Flow rate: 2 ml min^{-1}. Detector: absorbance monitor at 265 nm.

(4) Ion-exchange chromatography

Ion-exchange chromatography has rarely been used for the purification of ABA. On a preparative scale one problem would be the removal of involatile solvents. Sweetser and Vatvars (1976) used the pellicular cation exchanger Zipax-SCX with water, acidified to pH 1.7 with nitric acid as the mobile phase. The retention mechanism for ABA presumably involved adsorption and/or partition rather than ion-exchange.

Düring and Bachmann (1975) analysed ABA with the anion-exchanger AAX eluted with boric acid/sodium chlorate/water, pH 7. This is of limited

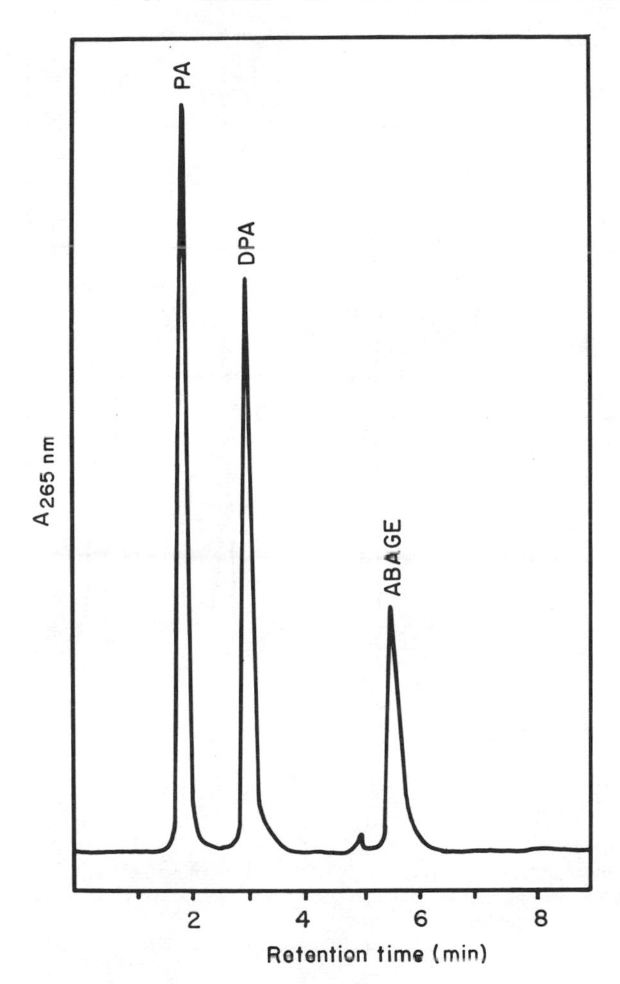

Fig. 7. Adsorption HPLC of ABA metabolites. Column: 250 × 4.6 mm i.d. 5 μm Partisil. Mobile phase: [chloroform/acetic acid (100:2, v/v)]/methanol (91:9, v/v). Flow rate: 2 ml min^{-1}. Detector: absorbance monitor at 265 nm.

value as an analytical tool as ABA is hardly retained at all, eluting just after the solvent front.

F. Gas chromatography

Gas chromatography (GC) has been the most extensively used technique for quantifying ABA. The detectors employed have been the flame-ionization

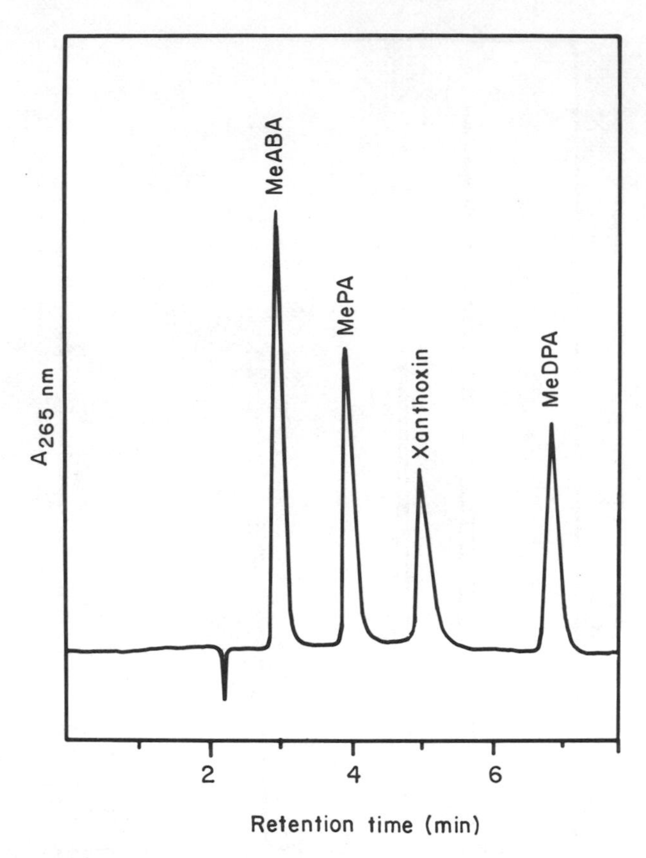

Fig. 8. Adsorption HPLC of MeABA and related compounds. Column: 250 × 4.6 mm i.d. 5 μm Spherisorb. Mobile phase: hexane/2-propanol (88:12, v/v). Flow rate 2 ml min^{-1}. Detector: absorbance monitor at 265 nm.

detector (FID) the ECD and the mass spectrometer. Their use is discussed in Section VIII.B. GC can also be a useful preparative tool. When operated in the preparative mode the column effluent is split by a T-piece before entering the detector, so that a portion of the effluent is diverted into a chilled trap.

ABA is not in itself sufficiently volatile for GC and therefore requires derivatization prior to analysis. MeABA is usually the derivative of choice. The most commonly employed methylating reagent diazomethane can be generated on a small scale by the method of Schlenk and Gillerman (1960) as shown in Fig. 11. Diazomethane is carcinogenic and explosive—generation of diazomethane and methylations should be performed in a fume cupboard behind safety glass. Small quantities (<20 ml) of diazomethane dissolved in

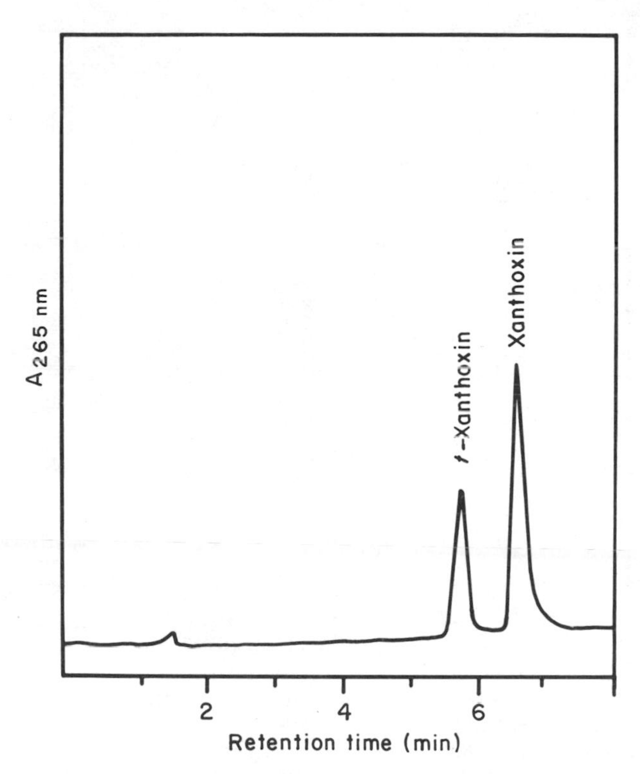

Fig. 9. Adsorption HPLC of xanthoxin. Column: 250 × 4.6 mm i.d. 5 μm Partisil. Mobile phase: chloroform/acetonitrile (95:5, v/v). Flow rate: 1 ml min^{-1}. Detector: absorbance monitor at 265 nm.

ether/methanol (4:1, v/v) can be stored at −20°C for extended periods. For methylation, dry samples should be dissolved in a small volume (*ca.* 1 ml) of diazomethane solution and left for *ca.* 10 min, during which time the yellow colour should be retained. If it rapidly disappears the diazomethane is exhausted and more should be added.

Whenham and Frazer (1981) have devised a simple radioassay for ABA analysis based on methylation with [^{14}C]diazomethane. The validity of this method is dependent upon the radiochemical purity of the purified [^{14}C]MeABA. The ethyl- and *p*-nitrobenzyl esters of ABA have also been used for GC analysis (Quarrie, 1978; Velaso *et al.*, 1978).

The methyl esters of PA and DPA have been analysed by GC (Harrison and Walton, 1975; Pierce and Raschke, 1981; Sivakumaran *et al.*, 1980; Tietz *et al.*, 1979) in addition to the 4′-TMS ether of MeDPA (**20**) (Martin *et al.*, 1977). The mass spectra of MePA, MeDPA and TMSMeDPA are shown in Appendix C.

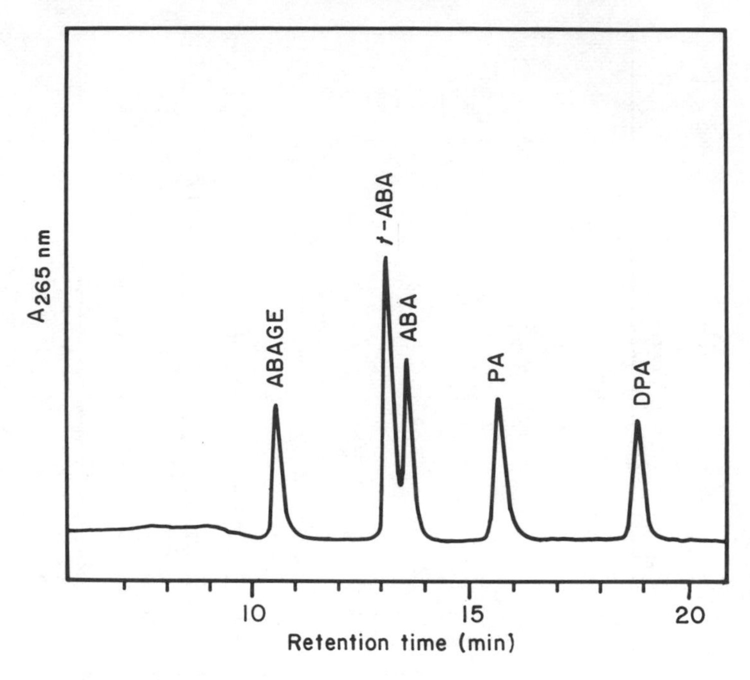

Fig. 10. Normal-phase HPLC of ABA and related compounds. Column: 250 × 4.6 mm i.d. 5 μm Partisil-PAC. Mobile phase: 30 min linear gradient of 0 to 20% aqueous 0.1 M acetic acid in acetonitrile. Flow rate: 2 ml min^{-1}. Detector: absorbance monitor at 265 nm.

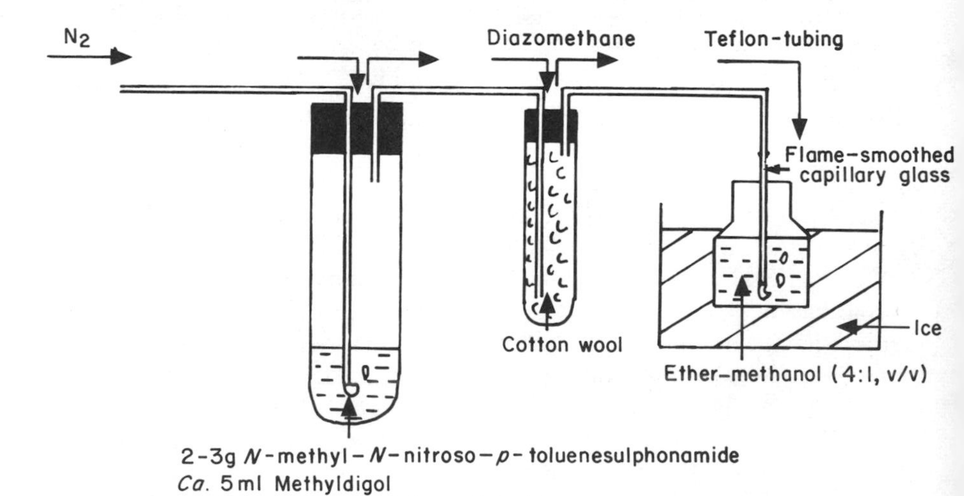

Fig. 11. Small-scale generation of diazomethane.

R_1O ··· OH ··· $O{=}C{-}OR_2$

$R_1 = Si(CH_3)_3$

$R_2 = CH_3$

20 Me TMS DPA

The tetra-acetylated methyl ester of DPAGS has been subjected to GC (Hirai and Koshimizu, 1983). Boyer and Zeevaart (1982) have chromatographed the tetra-acetyl derivative of ABAGE (**19**) on a short (460 × 2 mm) 3% SE-30 column, although our attempts have been unsuccessful.

Xanthoxin is sufficiently volatile to be subjected to GC without any derivatization provided that the column has been conditioned to remove adsorption sites (see below). However, it has usually been analysed as its *O*-acetyl derivative (**18**) (Firn *et al.*, 1972; Zeevaart, 1974). Acetylation is carried out with acetic anhydride and pyridine. After acetylation the *O*-acetyl derivative must be purified prior to GC. Acetylation of xanthoxin induces side-chain isomerization. A sample of mainly *cis*-xanthoxin was

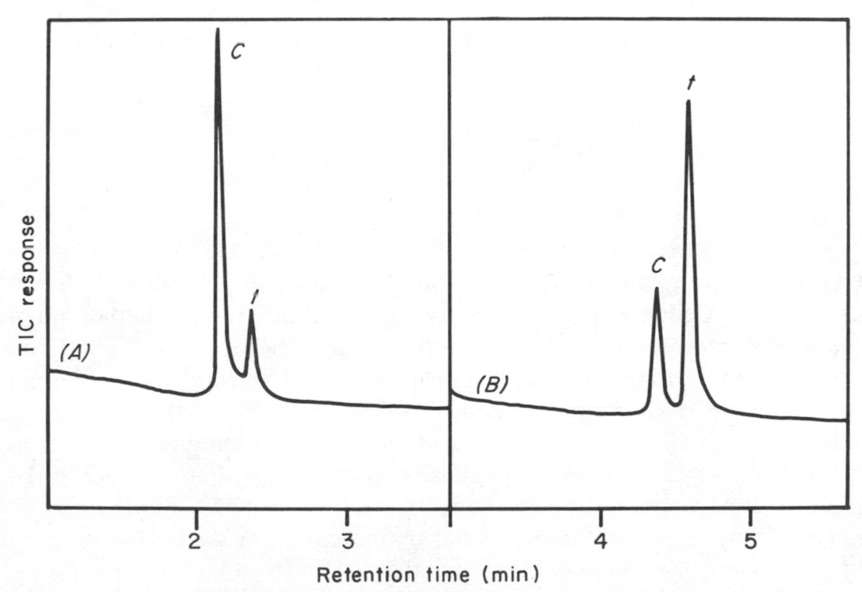

Fig. 12. Total ion current (TIC) capillary GC-MS traces of underivatized and acetylated xanthoxin. (*A*) Sample: xanthoxin. Column: 12 × 0.3 mm i.d. BP-1 WCOT (0.5 μm film). Column temperature: 163°C upwards at 6°C min^{-1}. Carrier gas: helium at 0.5 kg cm^{-2} (*B*) Sample: *O*-acetyl xanthoxin. Column and carrier gas as *A*. Column temperature: 150°C upwards at 6°C min^{-1}.

acetylated and purified by HPLC. Subsequent GC–MS analysis revealed that the *cis/trans* ratio had changed from 5:1 to 1:2.5 (Fig. 12), a phenomenon noted by other workers (see Dörffling and Tietz, 1983). This acetylation-induced isomerization means that any figures for endogenous *cis/trans* ratios must be viewed with scepticism. It is thus recommended that in any xanthoxin quantification procedures the free compound, not the *O*-acetyl derivative, be analysed. A further disadvantage of acetylation (or any other derivatization procedure) is that it may render volatile other compounds which, unlike xanthoxin, may not themselves be volatile enough to pass through the GC. The mass spectra of xanthoxin and *O*-acetyl xanthoxin are shown in Appendix C. Böttger (1978) has suggested the formation of a perfluorobutyryl derivative (**21**) so that the xanthoxin can be analysed with an ECD. Derivatization was carried out with perfluorobutyric acid anhydride, but apparently this procedure also induced side-chain isomerization. Taylor (quoted in Böttger, 1978) suggested reduction of the aldehyde group to the alcohol prior to derivatization to avoid isomerization.

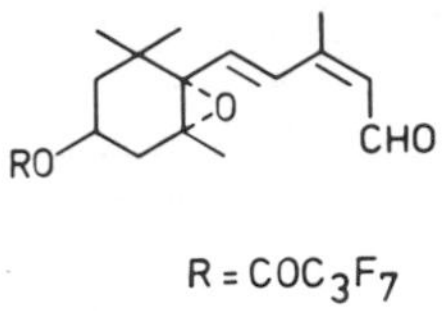

RO O CHO

$R = COC_3F_7$

21 *O*-heptafluorobutyryl xanthoxin

Numerous stationary phases have been used for the GC analysis of ABA and related compounds (Brenner, 1981; Dörffling and Tietz, 1983; Saunders, 1978). A good test of a stationary phase is to determine whether or not the two isomers of MeABA are well resolved without peak tailing and without unduly long analysis times. In analytical work at the nanogram level, column performance is critical. On older columns that have been conditioned at high temperatures we often find that column performance at the ng level is poor and peaks display extensive tailing. Performance at the μg level on the other hand is perfectly adequate. This situation can be remedied by a treatment consisting of injections of water followed by "Silyl-8" onto the column (Horgan and Neill, 1979). One explanation for this column behaviour and subsequent "cure" may be that the water hydrates residual adsorption sites on the packing material which can subsequently be inactivated by derivatization with the silylating reagent. After this treatment columns should be conditioned by several injections of relatively large quantities of MeABA to saturate any residual adsorption sites. During GC–MS analysis of ABA and [$^2H_3$]ABA, mixtures of labelled and unlabelled MeABA should be injected. This is because any adsorption sites may distinguish between the two species (see Millard, 1979).

Both non-polar and polar phases have been used for ABA analysis. Non-polar methyl silicone phases such as OV-1, SE-30, DC-200 and SP-2100 are commonly employed (Neill *et al.*, 1983; Gray *et al.*, 1979; Seeley and Powell, 1970; Harrison and Walton, 1975; Ciha *et al.*, 1977). These phases are very suitable for GC analysis of MeABA and related compounds especially when used for GC–MS, as they exhibit a low column "bleed". The GC separation of methylated ABA, PA and DPA is shown in Fig. 13. Some of the more polar stationary phases used include Epon-1001 (Saunders, 1978), XE-60 (Zeevaart and Milborrow, 1976) and Ultrabond-PEGS and Carbowax (Dumbroff *et al.*, 1983).

Capillary columns are increasingly being employed, particularly for GC–MS analysis. Wall-coated open tubular (WCOT) columns have a much higher efficiency than packed columns although, of course, a much reduced sample capacity. Because of their increased resolving power they can separate MeABA from contaminating materials where other columns may fail. Because of their high efficiency, peaks are very sharp and therefore the use of WCOT columns can increase the sensitivity of detection (see Section VIII.B). Capillary GC is likely to find even more use with the introduction of fused silica chemically bonded phase columns. We routinely use BP-1 fused quartz WCOT columns with a bonded non-polar stationary phase to analyse ABA and related compounds.

The type of injector used is a very important factor in the capillary GC analysis of MeABA. We have used both an on-column injector (SGE-OCI/1) and a Grob-type splitless injector (Carlo Erba) in conjunction with bonded-phase columns, for ABA analysis. Both types of injector have advantages and disadvantages. On-column injection guarantees the minimum of sample decomposition during injection, since injections are made by depositing the liquid sample directly onto the column at room temperature. The disadvantage of the method is that involatile impurities will collect at the point of injection and eventually build up to a point where column performance is impaired. Although bonded-phase columns may sometimes be cleaned by washing with an organic solvent, we have found that extensive use of an on-column injector with impure plant extracts results in loss of performance which can only be rectified by removing a metre or so of the column.

The above problem does not occur when splitless injection is used. However, due to the high injector temperature significant sample decomposition may occur if the injection system is not totally inert. Thus scrupulously

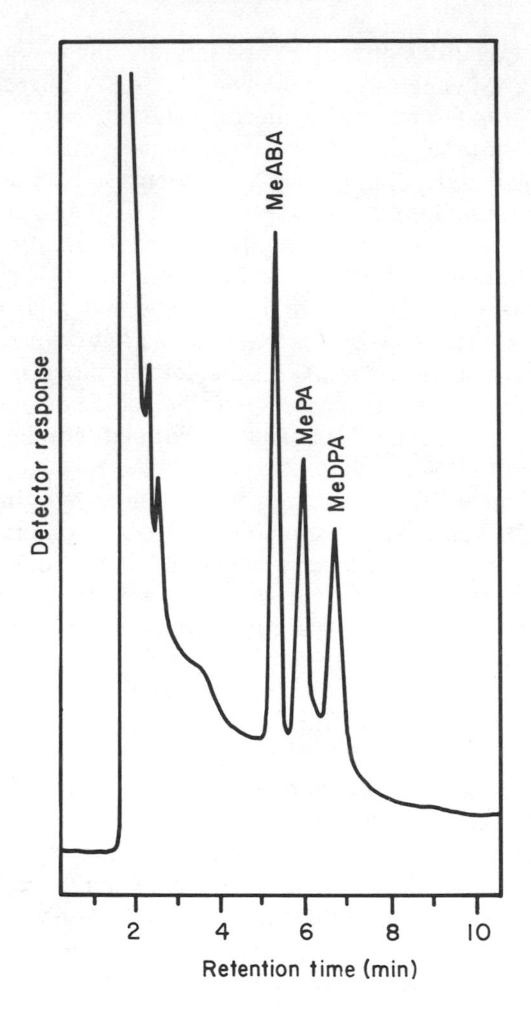

Fig. 13. GC of MeABA, MePA and MeDPA. Column: 1 m × 4 mm i.d. 3% OV-1 at 210°C. Carrier gas: nitrogen at 40 ml min^{-1}. Detector: ECD.

clean and well-silanized glass inserts should be used with this type of injector. Inserts should be changed as soon as a loss of performance is observed.

We have found that an equimolar mixture of methyl stearate, MeABA and tetracosane in ethyl acetate provides a good check on the performance of capillary columns and injectors. By injecting this mixture at 100 ng, 10 ng and 1 ng per component and observing the relative heights of the peaks, fall-off in system performance is easily observed.

VIII. QUANTIFICATION

A. High-performance liquid chromatography

Several types of HPLC detector are available, including refractive-index, fluorescence and UV monitors. Only the latter have been used for the quantification of ABA and are particularly suitable due to the high molar extinction coefficient of ABA. Many UV detectors are fixed at a particular wavelength, usually 254 nm, and the detection limit for ABA at this wavelength has been reported to be in the range 0.5–5 ng (see Brenner, 1981). The use of a variable wavelength detector tuned to the absorption maximum of the compound in question (e.g. 265 nm for MeABA) can increase the sensitivity. We have found that the level of detection is less than 5 ng but the linearity of response at the necessarily high sensitivities is lost.

Several workers have used reversed-phase HPLC with UV detection to quantify ABA. When ABA levels are high (and levels of other compounds low), as in *Cercospora* culture filtrates, this technique is reliable and reproducible (Griffin and Walton, 1982; Neill and Horgan, 1983; Norman *et al.*, 1982). However, reversed-phase HPLC has also been used to quantify ABA in crude plant extracts (e.g. Hardin and Stutte, 1981; Mousdale, 1981; Taylor and Rossall, 1982). In our experience UV detection is not suitable for quantitative analysis until a well-resolved single-component peak is obtained, i.e. after at least one HPLC step. In none of these studies was peak identity confirmed; we find that peaks co-eluting with ABA on reversed-phase HPLC of plant extracts are resolved into several components during subsequent chromatography.

Other workers have used adsorption and normal-phase HPLC to quantify ABA or its methyl ester after initial reversed-phase purification (Ciha *et al.*, 1977; Durley *et al.*, 1982; Wheaton and Bauscher, 1977; Cargile *et al.*, 1979). Only in the latter study was peak identity confirmed by GC–MS; however, the use of sequential HPLC reduces the chance of contamination. HPLC has also been used to quantify PA and DPA (Durley *et al.*, 1982; Hirai and Koshimizu, 1983). UV detection is less sensitive for these compounds due to their lower extinction coefficients.

No doubt new HPLC separation and specific detection techniques will appear in the next few years, and the use of quantitative HPLC will consequently increase. The lack of a requirement for derivatization and the capacity for large injection sizes give HPLC an advantage over GC as an analytical tool. However the higher limit of detection and the lack of selectivity of a UV-absorbance monitor compared to an ECD means that at present GC is still the preferred method for the analysis of ABA in the nanogram range.

In our experience, until more selective HPLC detectors become available, HPLC quantification of plant-derived ABA should be performed on samples previously purified by at least one other chromatographic step. Quantification techniques should be verified by confirmation of peak identity; thus the UV peak being quantified should be shown to contain ABA and no other UV-absorbing compounds. Several workers have demonstrated the presence of ABA in their samples by GC–MS; however, it does not necessarily follow that the peak being quantified by other means is due solely to ABA.

B. Gas chromatography

(1) Flame ionization detector

An FID will detect any compound that ionizes in a hydrogen flame. It is essentially non-selective and relatively insensitive, although FID has been used to quantify ABA in plant extracts (Lenton *et al.*, 1971; Hsu, 1979). Because of the lack of sensitivity, large samples or samples rich in ABA have to be used and extensive purification is required. As with UV detection the non-specific nature of an FID requires that peak identity should be established by other means. For example in one study ABA levels were apparently determined by GC–FID although the amounts of ABA injected on the column were below the detection limit (El-Antably, 1975). Xanthoxin has been quantified by FID (Firn *et al.*, 1972; Zeevaart, 1974); the same problems of large sample size and extensive purification are encountered. The use of capillary columns increases the sensitivity of an FID by sharpening the GC peaks and has the added advantage of better resolution from contaminating compounds.

(2) Electron capture detector

An ECD essentially consists of a radioactive source emitting β-electrons placed between two electrodes. By placing a potential difference across the electrodes an electric current is set up by flow of the emitted β-particles. Eluate from the GC column is swept through the detector; any electron-capturing compounds in the GC eluate attract some of the β-particles and hence lower the electron current. Early ECDs used ^{3}H as the electron source and monitored the decrease in electron current. Nowadays ^{63}Ni β-sources are usually used and the electron current is maintained at a constant value. An ECD is extremely sensitive compared to an FID, but only to electron-capturing molecules. Such molecules include halogenated compounds, certain aromatic hydrocarbons and ABA.

ECDs were first introduced for ABA analysis by Seeley and Powell (1970) and are now the most commonly used instruments for ABA analysis. An ECD has a wide dynamic range and sensitivity is over 1000 times that of a FID. With packed columns ABA can be quantified in the 0.05–10 ng range. Sensitivity can be extended down to 0.3 pg per injection by the use of capillary columns (Brenner, 1981). Because of the greatly increased selectivity, samples that are not suitable for FID analysis can often be analysed by ECD. Adequate sample purification is still necessary, however, especially as an ECD is easily overloaded and particularly prone to contamination. Peak identity should be confirmed for each tissue used, at least to show that the putative ABA peak actually contains ABA and no other electron-capturing compounds. As pointed out by Brenner (1981), rapid purification techniques such as that reported by Hubick and Reid (1980) must be viewed with caution.

ABA metabolites are also electron-capturing, although less so than ABA itself. Methylated PA and DPA, tetra-acetylated ABA-GE and acetylated methylated DPAGS have all been analysed by GC-ECD (Harrison and Walton, 1975; Pierce and Raschke, 1981; Zeevaart, 1983; Boyer and Zeevaart, 1982; Hirai and Koshimizu, 1983). Xanthoxin is not itself electron-capturing. Böttger (1978) derivatized xanthoxin to form the electron-capturing heptafluorobutyrate, the detection limit being less than 30 pg per injection. Although forming halogenated derivatives increases the sensitivity it does not increase the selectivity as any compounds in the extract with reactive groups are also derivatized.

(3) Mass spectrometry

MS linked to GC provides a highly sensitive and versatile detector system. Identification of the components in a sample is based on both GC retention time and mass spectral characteristics. Thus MeABA could be quantified by MS by monitoring the total ion current, and identity of the putative peak established directly. Usually *ca.* 10 ng of a compound is required for a full mass spectrum although with the use of more sensitive ionization techniques the sensitivity limit may be substantially lower.

In practice total ion current monitoring is not used to quantify MeABA as the much more sensitive and extremely selective technique of SIM is available. By focusing the instrument on a particular ion characteristic of MeABA the intensity of this ion as a function of GC retention time can be monitored, the resulting record being termed a single ion current profile or mass fragmentogram. As the ion current response is proportional to the actual amount of MeABA injected, suitable calibration curves can be constructed. The much improved signal/noise ratio of GC–SIM means that the sensitivity of GC–SIM compared to scanning MS is greatly increased, typically by a factor of over 1000.

Although GC–SIM offers very high selectivity it is not absolute, and other compounds present in extracts may give rise to the same ion as that being monitored: the chances of this occurring are greater at low molecular weights. Thus the highest molecular-weight ion possible, ideally the molecular ion, should be monitored. Unfortunately, intensities of the high molecular-weight ions in the EI–MS of MeABA are too low to be of value for GC–SIM. Consequently the base ion, at *m/z* 190, has usually been monitored (Andersson *et al.*, 1978; Browning, 1980; Hillman *et al.*, 1974). Andersson *et al.* (1978) found that the levels of ABA determined by GC–SIM were 10% higher than those determined by GC–ECD and suggested that this was due to interfering substances present in the extract that co-chromatographed with MeABA and also gave rise to an *m/z* 190 ion. Thus any SIM technique should be validated by demonstrating that only ions formed from MeABA contribute to the ion peak at *m/z* 190. GC–SIM has also been used to quantify PA and DPA (Sivakumaran *et al.*, 1980).

MIM is a variation of SIM. In MIM the mass spectrometer monitors more than one ion during a GC run by rapid switching of the accelerating voltage. Although this leads to a reduction in sensitivity it has the advantage that if two or more ions are monitored any change in their intensity ratio from that of a standard demonstrates the presence of impurities and thus provides a check on accuracy. Another advantage of MIM is that it enables heavy isotope labelled analogues to be used as internal standards. The use of MIM and ^{2}H-labelled ABA is discussed in Section IX.

Normally ions produced by electron impact have been monitored in SIM and MIM. However, CI–MS offers some advantages. As discussed earlier CI–MS produces high-molecular-weight ions in much greater intensities. Consequently, monitoring these ions should reduce the chances of interference from impurities.

C. Immunoassays

As was mentioned briefly in the introduction, the development of immunoassay methods for ABA and its derivatives represents an important milestone in the techniques of ABA analysis. Practical immunoassays for ABA were first described simultaneously by Walton *et al.* (1979) and Weiler (1979). Walton *et al.* used ABA linked through the carboxyl group to bovine serum albumin (BSA) to raise antibodies in rabbits, and Weiler used the same linking procedure to human serum albumin. Both procedures were radioimmunoassays using [^{3}H]ABA of high specific activity in competition for binding with the endogenous ABA. The results obtained by both workers were similar although Walton observed some anomalies between standard curves produced using (*RS*)-ABA and (*S*)-ABA. In both procedures antibodies were raised using racemic ABA and Walton pointed out the problems inherent in using these to determine naturally occurring ABA, which is exclusively (*S*).

The usable range of both assays was about 0.1–2.5 ng and cross-reactivity of PA and DPA, which would usually be present with ABA in crude extracts, was low. ABA-glucose ester tetraacetate and Me-ABA showed high (*ca* 100%) cross-reactivity. Results obtained by Walton *et al.* for the measurement of ABA levels in leaves of *Phaseolus vulgaris* and *Vicia faba* were comparable with those of independent measurements made by GC–ECD. Weiler surveyed a large number of plants for their ABA content and in general his results were in keeping with previously published work, although independent checks on ABA levels by an alternative method were not made.

Since it was clear that the assays described above, by virtue of the fact that the ABA was linked to the protein via its carboxyl group, could not discriminate between ABA and any carboxyl linked conjugates, Weiler (1980) has further developed the technique. He has produced an additional antiserum to ABA using ABA linked to BSA via the 4′-keto group with an intervening spacer group. This antiserum does not react with MeABA or ABA glucose ester tetraacetate, or with ABA glucose ester to any significant extent. At the same time a new antiserum to (*S*)-ABA was produced using carboxyl linking of (*S*)-ABA to BSA. Weiler has utilized both antisera in a differential technique to measure both ABA and carboxyl-conjugated ABA in a variety of plant extracts. Measurements were made on crude extracts and on TLC-purified material. In general the results, for tissues where levels had previously been studied by other workers using physical techniques, were comparable with published values. As with the previous technique, independent checks were not made. An outline of the procedures used to form carboxyl- and keto-linked ABA/BSA conjugates is shown in Fig. 14.

Details of a solid-phase enzyme-immunoassay (EIA) for ABA have been reported independently by Weiler (1982) and Daie and Wyse (1982) using ABA linked to alkaline phosphatase. Although this assay is somewhat less precise than the previously described RIAs it has advantages in speed, extended linear range and the fact that high-specific-activity [^{3}H]ABA may be used as an internal standard. Daie and Wyse noted some important differences between their EIA and the RIAs previously described by Weiler (1979, 1980). Most importantly they demonstrated that crude extracts could not be assayed directly with any degree of accuracy. Dilutions

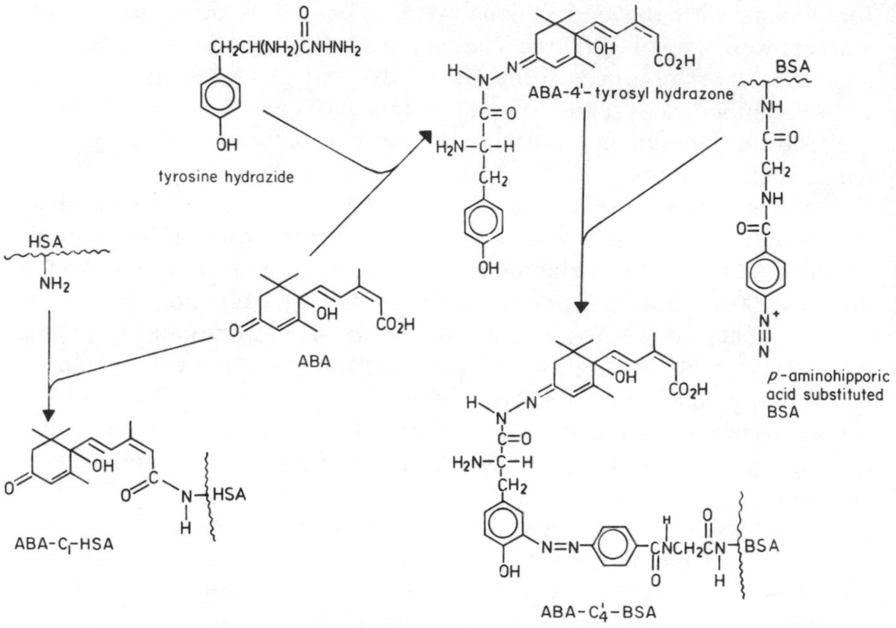

Fig. 14. Preparation of 1- and 4′-ABA-protein conjugates for immunoassay (Weiler, 1980).

of crude plant extracts did not show parallelism, thus indicating the presence of interfering compounds. It would appear, therefore, that some degree of purification is necessary before attempts at ABA measurement can be made by EIA.

Mertens *et al.* (1983) have further extended the immunoassay methods for ABA by developing an RIA using monoclonal antibodies raised against a 4′-keto linked ABA–BSA conjugate. Antibodies were obtained from two different cell clones and both exhibited extremely high affinities for (*S*)-ABA with *K* values of 7.9×10^9 and $3.7 \times 10^9\ \mathrm{l\,mol^{-1}}$. The measuring range of this assay is between 5 and 1000 pg of (S)-ABA. As would be predicted from the use of monoclonal antibodies the selectivity of this assay, with regard to the various ABA-related compounds tested, was even higher than that of the assays previously described. Yet again, no independent measurements of ABA have been made to confirm the accuracy of the technique.

An interesting extension of immunoassays for ABA has been the development of immunoaffinity absorbant for both (*R*)- and (*S*)-ABA by Mertens *et al.* (1982). These materials have been used to resolve high-specific-activity [^{3}H]-(*RS*)-ABA into apparently optically pure components. The capacity of these columns was too low to permit their use with plant extracts.

Knox and Galfre (1986) have described a column in which a monoclonal antibody to S-ABA was linked to Sepharose. A 5 ml column of this material could resolve 20 μg of (*RS*)-ABA. A preliminary investigation of the use of this material for purifying plant extracts (Neill and Knox, unpublished work), using GC–MS to assess sample purity, indicated that a very high degree of purification was obtainable with crude methanolic extracts of plant material. Thus it would appear that immunoaffinity purification of ABA may well become an important future technique.

It is clear from the brief description given above that the work described has led to the development of what are clearly extremely sensitive methods for the determination of ABA in plant extracts. An outline of the sensitivity and linear range of the various assays is shown in Table VI. The development of these techniques is a most impressive achievement and, if utilized wisely, they will provide plant physiologists with a major tool for the measurement of ABA levels in plant tissues leading eventually to an understanding of the subcellular distribution of ABA. Indeed, Sotta *et al.* (1985) claimed to have histochemically localized ABA in various organs of *Chenopodium polyspermum* L. using a rabbit antiserum to ABA and a soluble peroxidase–antiperoxidase complex.

As stated in the introduction, the authors have only limited practical experience of immunoassays for ABA, but nevertheless we feel that it would be appropriate to comment on the use of these techniques in relation to the other methods discussed in this chapter.

A major criticism that must be levelled at the work described above is that few independent checks have been made on the accuracy of immunoassay data. Rosher *et al.* (1985) have carried out a careful validation of an RIA for ABA using GC–MS as the standard measurement technique. For extracts of apple and sweet pepper tissues, they concluded that whilst the method was capable of good precision and accuracy, several important criteria had to be met. Because of the high sensitivity of the antiserum to ABA esters, prior

Table VI. Immunoassays for ABA.

Reference	Type of assay: Coupling via:	Usable range	Selectivity: cross-reactivity with: ABAGE; PA/DPA (50% displacement)
Walton *et al.* (1979)	RIA; carboxyl	0.1–2.5 ng	100; 3.6/0.6
Weiler (1979)	RIA; carboxyl	0.1–2.5 ng	103; 0.16/0.6
Weiler (1980)	RIA; carboxyl	0.05–8 ng	5.3; 0.5/0.08
	4′-keto	0.05–2 ng	0.2; 0.5/0.25
Mertens *et al.* (1983)	RIA (monoclonal) 4′ keto	0.005–1 ng	0; <0.1/<0.1
Weiler (1982)	EIA; carboxyl	0.03–80 ng	100; 0/0
Daie and Wyse (1982)	EIA; carboxyl	0.025–250 ng	?; 1.7/2

separation of free and conjugated ABA is required to prevent overestimation of the free ABA. In the case of apple extracts the use of PVP was found to reduce the amount of interference, although PVP was not required with sweet pepper extracts. The authors point out that although it was possible to obtain good precision with the plant materials investigated, this precision could be obtained only with the use of internal standards to monitor interference. This was particularly necessary in the case of apple extracts.

In contrast to the above findings, Neill and Horgan (unpublished work) have compared results obtained using an EIA based on a monoclonal antibody to (*S*)-ABA with those from GC–MS for crude and HPLC purified extracts of *Zea mays* kernels, tomato leaves and pine needles. In all cases EIA gave gross overestimates of ABA levels, particularly at high sample concentrations. These findings serve to emphasize the point that immunoassays have to be validated for each plant material and may well be unusable in certain circumstances. Although parallelism of extract dilutions and the standard curve may provide sound evidence for the absence of interference in these assays, proper checking of quantitative estimates by parallel measurements with a mass spectrometric technique, such as GC–MIM, would seem to be an obvious and essential prerequisite before immunoassay techniques can be fully accepted as reliable methods for ABA determination. In the light of the above results this would appear to be essential where crude plant extracts are to be used. It must also be pointed out that should some degree of purification be necessary, as when using antibodies binding both free and conjugated ABA, the inclusion of suitable internal standards is required for the reasons already stated in this chapter.

It would therefore seem appropriate to conclude this section on a cautionary note. Much of the confusion that exists with regard to past studies on the levels of endogenous plant growth regulators stems from the uncritical use of bioassays, and the interpretation of results derived from bioassays. Whilst it is unlikely that anything on this scale could occur with immunoassays, it will be most unfortunate if the full validity of the potentially very powerful immunoassay techniques for ABA is not properly established before they are put to widespread use.

IX. PURIFICATION LOSSES AND THE USE OF INTERNAL STANDARDS

With the exception of RIA (performed on crude extracts) all ABA quantification procedures involve extensive purification prior to analysis. Consequently, sample losses at each stage are unavoidable and can be considerable. These losses are due to adsorption on to glassware, incomplete recovery from chromatographic materials (e.g. silica gel), incomplete methylation, incomplete partitioning, poor dissolution, etc. Obviously these losses must be taken into account when making quantitative estimates.

Losses can be assessed by determining the recovery of an internal standard added at the extraction stage. The internal standard must be as chemically similar to ABA as possible, so that it is not separated from ABA during work-up, yet the final analytical step must be able to distinguish it from endogenous ABA.

Sample losses vary from extract to extract so that ABA recovery must be assessed for each sample. Some losses are not proportional to the ABA concentration in a sample. For example, once all of the adsorption sites in a system are filled then no more ABA will be lost through adsorption. Consequently, sample losses in extracts containing large amounts of ABA will be less than in extracts low in ABA. In our experience ABA recoveries can be 50% and higher in large extracts (e.g. 10–20 g fresh weight) yet less than 20% in extracts of less than 2 g. Variation in recovery between samples can be considerable. Several workers have added [^{14}C]ABA to just one or two extracts, determined the percentage recovery, and assumed recovery for all their other samples to be the same. This is unlikely to be the case. Other workers have added a large quantity of non-labelled ABA to an extract to determine recovery and again assumed the recovery to be the same for all samples. Thus their recovery figures for other samples are likely to be inaccurate overestimates, as well as being imprecise due to sample variation.

Milborrow (1967) described an elegant "racemate dilution" technique to assess recovery. Synthetic (*RS*)-ABA was added to extracts and the purified ABA analyzed by UV and ORD spectrometry. Thus the quantity of synthetic and endogenous ABA in the final sample could be determined. Due to its relative insensitivity this technique has found little use (see Section V).

Lenton *et al.* (1971) recommended the use of *t*-ABA as an internal standard. This is a perfectly satisfactory technique provided that no isomerization of the natural ABA occurs and that the purification procedure does not discriminate against either of the two isomers. *t*-ABA is not often used these days, as modern chromatographic procedures readily resolve the two isomers. However, it was used as an internal standard during quantification of ABA in crude *Cercospora* culture filtrates (Neill and Horgan, 1983). The use of *t*-ABA had the advantage that it could also be used as an injection standard (see below).

Usually radioactive ABA is added as an internal standard, and both [^{14}C]- and [^{3}H]ABA have been used (Little *et al.*, 1978; Walton *et al.*, 1979). A known amount of radioactive ABA is added at the extraction stage and the amount of radioactivity remaining after analysis determined. Percentage recovery is simply determined as the ratio of recovered radioactivity to added radioactivity. Assessment of recovery is dependent on the radiochemical purity of the recovered ABA. Although [^{14}C]- and [^{3}H]ABA are chemically stable, errors may possibly arise through incomplete methylation, etc. If the MeABA was not purified prior to analysis then the radioactivity determined would represent both MeABA and underivatized ABA, i.e. there would be an overestimate.

Another problem with radioactive standards can occur when small tissue samples are analysed. If only a few nanograms of endogenous ABA are present and an internal standard of low-specific-activity ABA added then the quantity of standard may be far greater than that of endogenous ABA. Consequently errors associated with subtracting the amount of standard ABA added from the final figure may be large. The synthesis of [^{3}H]ABA of very high specific activity (Walton *et al.*, 1977) has removed this problem.

The method of choice for quantifying ABA and simultaneously accounting for recovery losses involves the addition of heavy isotope-labelled internal standards and quantification by GC–MIM. Deuterium-labelled ABA has been used for this purpose. Polydeuterated standards are preferred, since ions from these will be distinct from those occurring in endogenous compounds through natural isotopic abundances. Rivier *et al.* (1977) used hexadeuterated ABA as an internal standard and quantified MeABA by GC–MIM using EI. Ions at *m/z* 190 and *m/z* 194 were monitored, being respectively the base peaks of endogenous and labelled ABA. This method has the disadvantage that the hexadeuterated ABA is synthesized by base-induced hydrogen exchange. Thus the pH of the extract must be kept acid to avoid re-exchange. We have used trideuterated ABA, labelled with three deuterium atoms in the side-chain methyl group (Horgan, 1981; Neill *et al.*, 1983). The label is introduced into a synthetic precursor via base-induced exchange and is rendered stable by subsequent chemical reaction (see Appendix B). We have used EI–MS, monitoring the ions at *m/z* 190 and 193.

Netting *et al.* (1982) and Rajasekaran *et al.* (1982) have also used 3-methyl-[^{2}H]ABA as an internal standard for MIM. However, they used methane CI–MS to ionize MeABA. CI–MS has the advantage that higher molecular-weight ions (in this case MH^{+}-H_2O) can be monitored, thus reducing the risk of interference.

[^{2}H]ABA is added to the extract and after purification ions at *m/z* 190 and 193 monitored by GC–MIM. In theory the ratio of response of unlabelled to labelled MeABA multiplied by the amount of labelled ABA standard added equals the amount of endogenous ABA originally present in the extract. In practice however, a standard curve relating $^1H/^2H$ peak area (or height) to $^1H/^2H$ injected must be constructed (see Fig. 15). This is because of the isotopic composition of the deuterated ABA.

Peak area ratio is determined by the expression

$$R = \frac{E + L'}{E' + L}$$

where R is the peak area ratio, E is the response at m/z 190 from unlabelled, E' is the response at m/z 193 from unlabelled, L' is the response at m/z 190 from labelled and L is the response at m/z 193 from labelled.

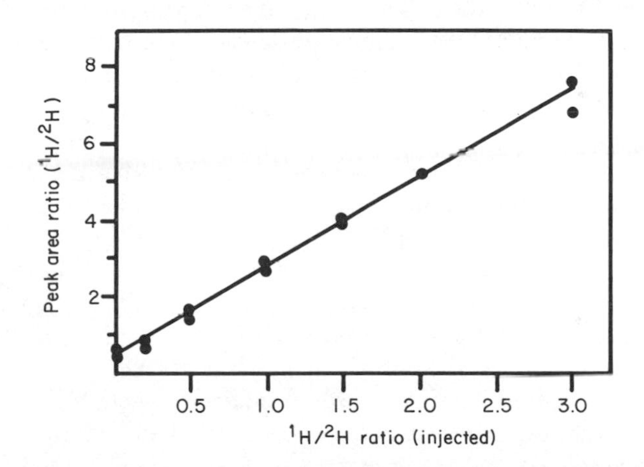

Fig. 15. GC–MIM calibration curve for isotopic dilution analysis of ABA obtained by co-injecting 5 ng[^{2}H]ABA with 0–15 ng[^{1}H]ABA.

The graph does not pass through zero because of the presence of an ion at m/z 190 in the labelled material. At low concentrations of [^{1}H]MeABA the graph is linear as E' tends to zero and L and L' are constant. The gradient is 2.3 not 1 because of the isotopic composition of the deuterium-labelled material ([3× ^{2}H]:[2 × ^{2}H]:[1 × ^{2}H]:[0 × ^{2}H]; 0.45: 0.32:0.15: 0.08). At very high concentrations of [^{1}H]MeABA the graph departs from linearity as E' becomes significant. Consequently the efficiency of the MIM technique is maximized when the amounts of endogenous ABA and internal standard are within the same range.

The corrected $^1H/^2H$ ratio multiplied by the amount of internal standard added gives the quantity of endogenous material present in the extract. Theoretically the only errors in MIM analysis are those involved in dispensing the deuterated internal standard, as injection errors and errors due to machine sensitivity are accounted for by the presence of the internal standard. Errors in making up solutions are likely to be greater than those involved in the mass spectrometric determinations (Millard, 1979). Thus when constructing calibration curves replicate samples should be made up and analysed singly, rather than by performing replicate analyses of a single stock solution.

A Other sources of error

If the detector response at the retention time of ABA (or MeABA) is not due solely to ABA but also to the presence of contaminating materials then the results will be inaccurate. The purity of an ABA chromatographic peak must be established, or at least it must be demonstrated that ABA is the only compound giving rise to a signal in the particular detector (e.g. ECD or SIM). This is probably the source of the bulk of the errors associated with ABA analysis.

The purity of internal standards must be checked before use, or else a constant non-random error will be introduced. Any pipetting errors introduced here will be reflected in the final analysis.

Injection errors and variations in machine sensitivity at the quantitation stage can introduce significant discrepancies into the final results. Injection errors are probably greater in GC than in HPLC as the injection volume is so much smaller relative to the sample size. The best method to account for injection and machine error is to use an internal injection standard. Calibration curves should be constructed in order to check that the peak-height ratios between ABA and the injection standard remain constant.

t-ABA and androsta-1,4-dien-3,17-dione have been used as internal standards for HPLC (Neill and Horgan, 1983; Cargile *et al.*, 1979) and ethyl-ABA and tetraphenyl ethylene have been used for GC analysis (Quarrie, 1978; Hsu, 1979).

APPENDIX

A. Quantification of abscisic acid

The following procedures are two "recipes" for ABA quantification in plant tissues based on (1) TLC purification and quantification by GC–ECD, losses being accounted for by addition of high specific activity [^{3}H]ABA; (2) HPLC purification and quantification by GC–MIM, losses being assessed by inclusion of a [^{2}H]ABA internal standard. Obviously the two recipes are interchangeable, and GC–ECD quantification can follow HPLC purification. The procedures will require modification according to the quantity and nature of the tissue being extracted. For example, the second chromatographic steps or solvent partitioning may not be necessary. Conversely it may be necessary to perform a basic ether extraction or PVP purification if the extracts are particularly large.

Homogenize 5–10 g fresh weight of frozen or fresh tissue in 80% acetone/0.1 M acetic acid (5–10 ml g^{-1}).

Add the internal standard: (a) a known amount of high specific-activity [^{3}H]ABA (*ca.* 20 000 cpm); (b) a known amount of [^{2}H]ABA (add approximately the same quantity as the amount of endogenous ABA expected to be in the extract. This can be estimated by a preliminary extraction using [^{3}H]ABA as internal standard or alternatively with no internal standard).

Filter and reduce to aqueous *in vacuo*.

Freeze, thaw and centrifuge. Bench centrifugation is usually adequate to remove suspended materials.

Adjust pH to *ca.* 3. Extract three times with an equal volume of ether.

Reduce organic phase to dryness *in vacuo*.

(1) TLC GC–ECD

Dissolve extract in *ca.* 100 μl methanol. Streak onto 0.5 mm F_{254} silica gel TLC plate. Develop in toluene/ethyl acetate/acetic acid (50:30:4, v/v).

Elute opposite ABA marker spot with 1–2 ml water-saturated ethyl acetate, reduce to dryness.

Methylate with <1 ml diazomethane solution in ether/methanol (4:1, v/v).

Re-TLC in hexane/ethyl acetate (1:1, v/v).

Elute opposite MeABA marker spot. Dry in small vial.

Add known amount of *t*-MeABA as GC standard. Add an amount approximately equal to the amount of MeABA in the sample (estimated from previous extractions or by preliminary GC determinations).

Quantify by GC–ECD. Calculated amount of ABA in final sample from

$$A = \frac{c}{t} \times R \times T$$

where A is the amount of endogenous MeABA in the sample, c is the peak height of MeABA, t is the peak height of *t*-MeABA, R is the *trans*/*cis* peak-height ratio of equimolar standard mixture and T = amount of *t*-MeABA added to sample.

Remove an aliquot from the sample for scintillation counting to determine percentage recovery. Correct A for purification losses and thus calculate initial ABA concentration in extracted tissue.

(2) HPLC–GC–MIM

Dissolve the extract in 1 ml 20% methanol/0.1 M aqueous acetic acid. Filter through glass microfibre.

Fractionate by reversed-phase HPLC on a semipreparative (10 mm i.d.) ODS-column. Collect eluate around the retention time of ABA (previously determined by injection of a standard). (See Fig. 16).

Reduce to dryness. Methylate with diazomethane solution in ether/methanol (4:1).

Dissolve in *ca.* 250 μl of hexane/2-propanol (95:5, v/v). Purify methyl ester by adsorption HPLC on silica gel column in same solvent. (See Fig. 17.)

If the dry weight of the extract following reversed-phase HPLC is too large then methylation efficiency may be impaired or, more likely, subsequent dissolution in the hexane solvent is reduced. Consequently final recoveries may be unacceptably low. In these cases it may be preferable for the second chromatographic step to involve analytical reversed-phase HPLC of the methyl ester or adsorption or normal-phase HPLC of the free acid using a stronger solvent such as chloroform/acetic acid (100:2, v/v).

Collect MeABA fraction and blow dry in small vial for subsequent GC–MIM analysis.

Quantify by GC–MIM. Calculate peak-area ratio (unlabelled/labelled) and obtain corrected ratio value from calibration curve. The corrected ratio multiplied by the

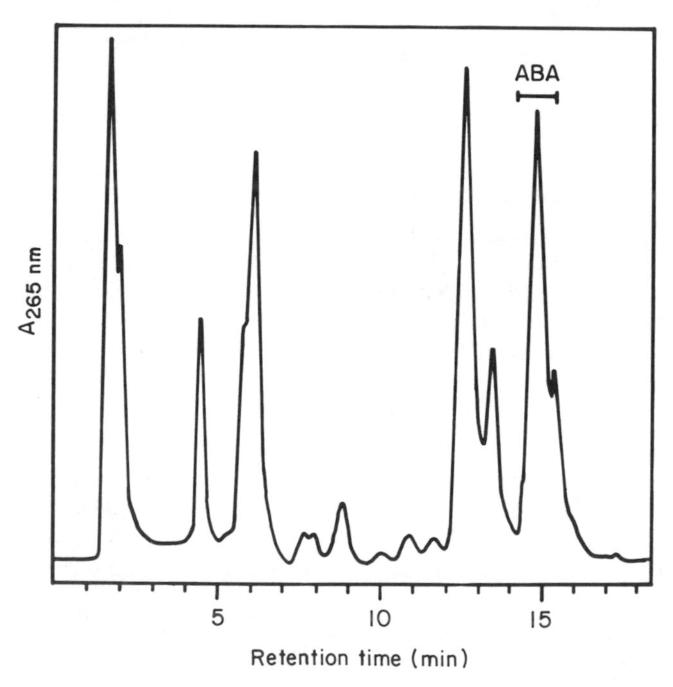

Fig. 16. Reversed-phase HPLC of a semi-purified extract from 5 g (f.w.) of bean leaves Column: 150 × 10 mm i.d. 5 μm ODS-Hypersil. Mobile phase: 40 min linear gradient of 20 to 100% methanol in 0.1 M aqueous acetic acid. Flow rate: 5 ml min^{-1}. Detector: absorbance monitor at 265 nm.

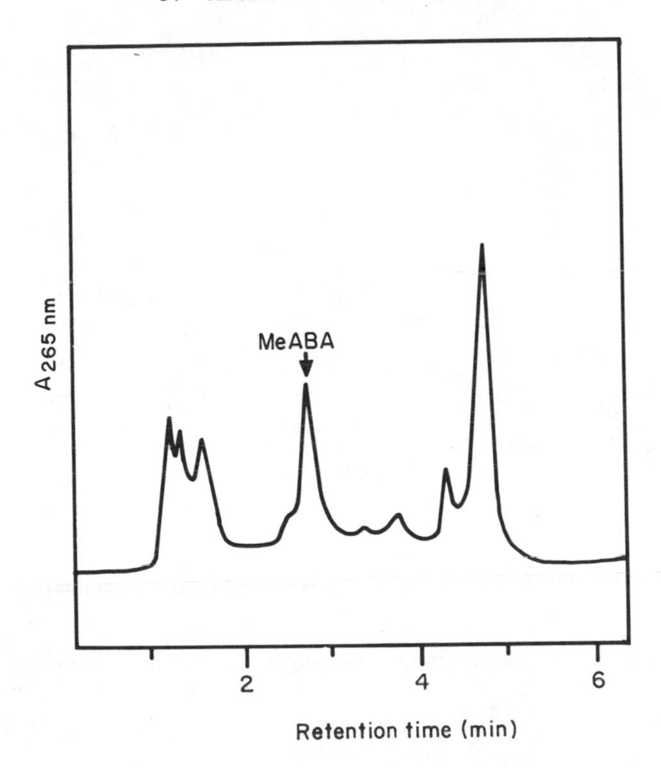

Fig. 17. Adsorption HPLC of the ABA fraction from Fig. 16 following methylation. Column: 250 × 4.6 mm i.d. 5 μm Spherisorb. Mobile phase: hexane/iso-propanol (95:5, v/v). Flow rate: 2 ml min^{-1}. Detector: absorbance monitor at 265 nm. Note that the large UV-absorbing peak at the ABA retention time in Fig. 16 is impure as it contains a number of components in addition to ABA.

amount of labelled standard added equals the amount of endogenous ABA initially present in extract.

B. Preparation of [^{2}H]-labelled ABA for use as an internal standard

The method described below and illustrated in Fig. 18 is based on the exchange of the enolizable hydrogens of the hydroxydiketone (**22**) with [^{2}H]H_2O at high pH. The deuterated diketone is converted to ABA by the standard Wittig procedure. The deuterium atoms incorporated into the side-chain methyl group of ABA-ethyl esters thus formed are stable whilst the other deuterium atoms re-exchange during the subsequent hydrolysis step.

100 mg of hydroxydiketone prepared by the method of Roberts *et al.* (1968) was dissolved in 0.8 ml of dry dioxane and 0.4 ml of [^{2}H]H_2O was added. 5 mg of freshly

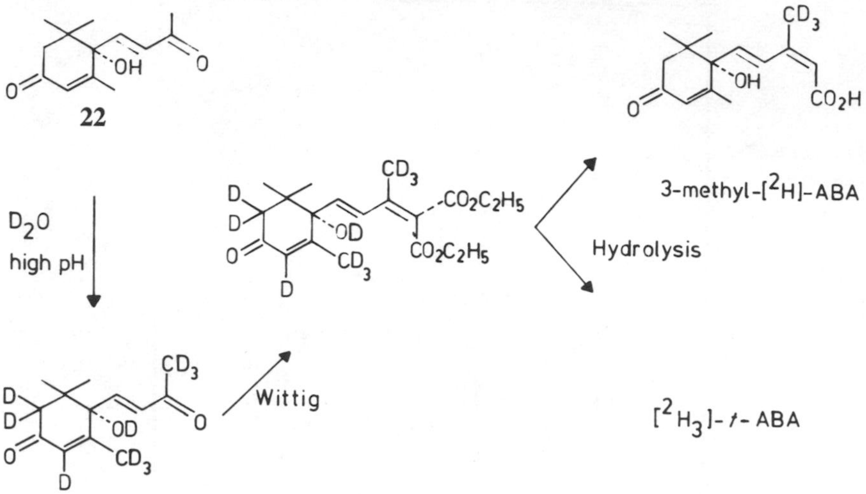

Fig. 18. Preparation of 3-methyl-[$^2H_3$]ABA.

cut sodium was added carefully and the mixture was heated in a sealed tube at 100°C for 10 min. After standing at room temperature overnight the tube was opened and the reaction mixture poured into a 100 ml conical flask containing 75 ml of dry diethyl ether and 15 g of Drierite. The flask was stoppered and allowed to stand at 4°C for 48 h. The Drierite was removed by filtration and the filtrate evaporated to dryness to yield 72 mg of crude deuterated hydroxydiketone.

A solution of 150 mg of carboethoxymethylene triphenyl phosphorane in 1 ml of dry ethyl acetate was added to the crude hydroxydiketone. After the mixture had dissolved completely the ethyl acetate was removed with a stream of dry nitrogen. The residue was heated under nitrogen on an oil bath for 1 h while the temperature was slowly raised from 150 to 170°C. After cooling, the mixture was triturated with 1 ml of diethyl ether and allowed to stand at −20°C for 1 h. The precipitate of triphenyl phosphine oxide was removed by vacuum filtration and washed once with 1 ml of cold diethyl ether. The filtrate was evaporated to dryness and redissolved in 0.8 ml of methanol. 0.5 ml of 5 M potassium hydroxide was added and the mixture allowed to stand at room temperature for 12 h. 10 ml of water was added and the mixture extracted with 10 ml of diethyl ether which was discarded. The aqueous residue was acidified to pH 2.5 with 5 N hydrochloric acid and extracted with three 10 ml portions of diethyl ether.

The combined ether extracts were dried over magnesium sulphate to yield, after evaporation, 70 mg of a crude mixture of [$^2H_3$]ABA and [$^2H_3$]*t*-ABA. Pure ABA was isolated from this crude product as required by TLC or HPLC and used for the determination of ABA in plant extracts as described in the text. GC–MS of the purified ABA as MeABA showed a molecular composition of [$^2H_3$:$^2H_2$:$^2H_1$:$^2H_0$, 0.45:0.32:0.15:0.08] measured in the base peak.

C. 10-peak mass spectra of ABA and related compounds

EI–MS of MeABA (70 eV; GC–MS)
M^+278 (3), 260 (5), 246 (5), 191 (18), b.p. 190 (100), 162 (30), 134 (29), 125 (42), 112 (11), 91 (14).

CI–MS of MeABA (methane; GC–MS) (Netting *et al.*, 1982)
319 (5), 307 (11), $[MH]^+$ 279 (15), 262 (16), b.p. 261 (100), 247 (43), 229 (14), 219 (9), 190 (6), 125 (9).

CI–MS of MeABA (iso-butane; GC–MS) (Rivier and Saugy, 1986)
$[MH]^+$ 279 (66), 263 (40), b.p. 261 (100), 247 (20), 190 (4), 149 (12).

CI–MS of MeABA (ammonia; GC–MS) (Rivier and Saugy, 1986)
b.p. 296 (100), $[MH]^+$ 279 (58), 261 (60), 190 (8), 125 (4).

Negative ion CI–MS of MeABA (methane; GC–MS) (Rivier and Saugy, 1986)
$[M]^-$ b.p. 278 (100), 262 (4), 260 (22), 246 (5).

Negative ion CI–MS of MeABA (iso-butane; GC–MS) (Rivier and Saugy, 1986)
$[M]^-$ b.p. 278 (100), 263 (15), 260 (33), 246 (10).

Negative ion CI–MS of MeABA (ammonia; GC–MS) (Rivier and Saugy, 1986)
$[M]^-$ b.p. 278 (100), 260 (28), 245 (5).

EI–MS of MePA (70 eV; GC–MS)
M^+ 294 (10), 167 (34), 154 (33), 139 (39), 135 (38), b.p. 125 (100), 122 (78), 121 (72), 109 (34), 94 (45).

EI–MS of MeDPA (70 eV; GC–MS)
M^+ 296 (5), 278 (21), 154 (52), 127 (47), 125 (48), b.p. 122 (100), 121 (56), 109 (58), 95 (48), 94 (47).

EI–MS of MeTMsDPA (24 eV, GC–MS) (Martin *et al.*, 1977)
M^+ 368 (1), 220 (14), 177 (46), 159 (76), 125 (44), 122 (50), 121 (56), 109 (45), 75 (41), 73 (100).

EI–MS of tetra-acetylated ABAGE (70 eV, probe)
441 (2), 331 (70), 271 (15), 247 (22), 190 (38), b.p. 169 (100), 145 (15), 127 (20), 109 (65).

CI–MS of tetra-acetylated ABAGE (ammonia; probe)
612 (9), $[M + H]^+$ 595 (4), 460 (6), b.p. 366 (100), 331 (75), 306 (21), 246 (38), 213 (35), 186 (20).

EI–MS of xanthoxin (70 eV; GC–MS)
M^+ 250 (11), 175 (4), 160 (18), 149 (38), 133 (15), 123 (24), 121 (22), 107 (23), 95 (44), 43 (100).

EI–MS of O-acetyl xanthoxin (70 eV; GC–MS)
M^+ 292 (2), 232 (20), 176 (21), 161 (11), 149 (100), 133 (37), 121 (22), 105 (25), 95 (22), 43 (100).

REFERENCES

Addicott, F. T. (1983). "Abscisic Acid." Praeger, New York.

Andersson, B., Haggstrom, N. and Andersson, K. (1978). Identification of abscisic acid in shoots of *Picea abies* and *Pinus sylvestris* by combined gas chromatography–mass spectrometry. A versatile method for cleanup and quantification. *J. Chromatogr.* **157**, 303–310.

Bangerth, F. (1982). Changes in the ratio of *cis,trans* to *trans,trans* abscisic acid during ripening of apple fruits. *Planta* **155**, 199–203.

Beardsell, M. F. and Cohen, D. (1975). Relationship between leaf water status, abscisic acid levels and stomatal resistance in maize and *Sorghum*. *Plant Physiol.* **56**, 207–212.

Böttger, M. (1978). The occurrence of *cis,trans* and *trans,trans*-xanthoxin in pea roots. *Z. Pflanzenphysiol.* **86**, 265–268.

Boyer, G. L. and Zeevaart, J. A. D. (1982). Isolation and quantitation of β-D-glucopyranosyl abscisate from leaves of *Xanthium* and spinach. *Plant Physiol.* **70**, 227–231.

Brenner, M. L. (1981). Modern methods for plant growth substance analysis. *Annu. Rev. Plant Physiol.* **32**, 511–538.

Browning, G. (1980). Endogenous *cis,trans*-abscisic acid and pea seed development: evidence for a role in seed growth from changes induced by temperature. *J. Exp. Bot.* **31**, 185–197.

Cargile, N. L., Borchert, R. and McChesney, J. D. (1979). Analysis of abscisic acid by high performance liquid chromatography. *Anal. Biochem.* **97**, 331–339.

Ciha, A. J., Brenner, M. L. and Brun, W. A. (1977). Rapid separation and quantification of abscisic acid from plant tissues using high performance liquid chromatography. *Plant Physiol.* **59**, 821–826.

Cornforth, J. W., Draber, W., Milborrow, B. V. and Ryback, G. (1967). Absolute stereochemistry of (±)-abscisin II. *Chem. Commun.* 114–116.

Cornforth, J. W., Mallaby, R. and Ryback, G. (1968). Synthesis of (±)-abscisin II. *Nature* **206**, 715.

Creelman, R. A. and Zeevaart, J. A. D. (1984). Incorporation of oxygen into abscisic acid and phaseic acid from molecular oxygen. *Plant Physiol.* **75**, 166–169.

Daie, J. and Wyse, R. (1982). Adaptation of the enzyme-linked immunosorbent assay (ELISA) to the quantitative analysis of abscisic acid. *Anal. Biochem.* **119**, 365–571.

Dathe, W. and Sembdner, G. (1982). Isolation of 4′-dihydroabscisic acid from immature seeds of *Vicia faba*. *Phytochemistry* **21**, 1798–1799.

Dörffling, K. and Tietz, D. (1983). Methods for the detection and estimation of abscisic acid and related compounds. *In* "Abscisic Acid" (F. T. Addicott, ed.) pp. 23–77. Praeger, New York.

Dumbroff, E. B., Walker, M. A. and Dumbroff, P. A. (1983). Choice of methods for the determination of abscisic acid in plant tissue. *J. Chromatogr.* **256**, 439–446.

Düring, H. and Bachmann, O. (1975). Abscisic acid analysis in *Vitis vinifera* in the period of endogenous bud dormancy by high pressure liquid chromatography. *Physiol. Plant.* **34**, 201–203.

Durley, R. C., Kannangara, T. and Simpson, G. (1978). Analysis of abscisins and 3-indolylacetic acid in leaves of *Sorghum bicolor* by high performance liquid chromatography. *Can. J. Bot.* **56**, 171–176.

Durley, R. C., Kannangara, T. and Simpson, G. (1982). Leaf analysis for abscisic, phaseic and 3-indolylacetic acids by high performance liquid chromatography. *J. Chromatogr.* **236**, 181–188.

El-Antably, H. (1975). Re-distribution of indole acetic acid, abscisic acid and gibberellins in geotropically stimulated *Ribes nigrum* roots. *Z. Pflanzenphysiol.* **75**, 17–24.

Findlay, J. A. and Mackay, W. D. (1971). Synthesis of (RS)-abscisic acid. *Can. J. Chem.* **49**, 2369–2371.

Firn, R. D., Burden, R. S. and Taylor, H. F. (1972). The detection and estimation of the growth inhibitor xanthoxin in plants. *Planta* **102**, 115–126.

Glenn, J. L., Kuo, C. C., Durley, R. C. and Pharis, R. P. (1972). Use of insoluble polyvinylpyrrolidone for purification of plant extracts and chromatography of plant hormones. *Phytochemistry* **11**, 345–351.

Gray, R. T., Mallaby, R., Ryback, G. and Williams, V. P. (1974). Mass spectra of methyl abscisate and isotopically-labelled analogues. *J. Chem. Soc. Perkin Trans.* **2**, 919–924.

Griffin, D. H. and Walton, D. C. (1982). Regulation of abscisic acid formation in *Mycosphaerella* (*Cercospora*) *rosicola* by phosphate. *Mycologia* **74**, 614–618.

Hardin, J. M. and Stutte, C. A. (1981). Analysis of plant hormones using high-performance liquid chromatogaphy. *J. Chromatogr.* **208**, 124–128.

Harrison, M. A. and Walton, D. C. (1975). Abscisic acid metabolism in water-stressed bean leaves. *Plant Physiol.* **56**, 250–254.

Hillman, J. R., Young, I. and Knights, B. A. (1974). Abscisic acid in leaves of *Hedera helix*. *Planta* **119**, 263–266.

Hirai, N. and Koshimizu, K. (1983). A new conjugate of dihydrophaseic acid from avocado fruit. *Agric. Biol. Chem.* **47**, 365–371.

Hirai, N., Fukui, H. and Koshimizu, K. (1978). A novel abscisic acid metabolite from seeds of *Robinia pseudocacia*. *Phytochemistry* **17**, 1625–1627.

Hiron, R. W. P. and Wright, S. T. C. (1973). The role of endogenous abscisic acid in the response of plants to stress. *J. Exp. Bot.* **21**, 769–781.

Horgan, R. (1981). Modern methods for plant hormone analysis. *In* "Progress in Phytochemistry Vol. 7" (L. Rheinhold, J. B. Harborne and T. Swain, eds.), pp. 137–170. Pergamon, Oxford.

Horgan, R. and Neill, S. J. (1979). Column conditioning for trace analysis of abscisic acid by gas chromatography. *J. Chromatogr.* **177**, 116–117.

Horgan, R., Neill, S. J., Walton, D. C. and Griffin, D. (1983). Biosynthesis of abscisic acid. *Trans. Biochem. Soc.* **11**, 553–557.

Hsu, F. C. (1979). Abscisic acid accumulation in developing seeds of *Phaseolus vulgaris* L. *Plant Physiol.* **63**, 552–556.

Hubick, K. T. and Reid, D. M. (1980). A rapid method for the extraction and analysis of abscisic acid from plant tissue. *Plant Physiol.* **65**, 523–525.

Kienzle, F., Mayer, H., Minder, R. D., Thommen, H. (1978). Synthese von optisch aktiven, natürlichen Carotenoiden und Strukturell verwardten Verbindungen. III.

Synthese von (+)-Abscisinsaüre, (−)-xanthoxin, (−)-Loliolid, (−)-Actinidiolid und (−)-Dihydroactinidiolid. *Helv. Chim. Acta* **61**, 2616–2627.

Knox, J. P. and Galfre, G. (1986). Use of monoclonal antibodies to separate the enantiomers of abscisic acid. *Anal. Biochem.* **155**, 92–94.

Koshimizu, K., Inui, M., Fukui, H. and Mitsui, T. (1968). Isolation of (+)-abscisyl-β-D-glucopyranoside from immature fruit of *Lupinus luteus*. *Agric. Biol. Chem.* **32**, 789–791.

Lehmann, H., Preiss, A. and Schmidt, J. (1983). A novel abscisic acid metabolite from cell suspension cultures of *Nigella damascena*. *Phytochemistry* **22**, 1277–1278.

Lenton, J. R., Perry, V. M. and Saunders, P. F. (1971). The identification and quantitative analysis of abscisic acid in plant extracts by gas-liquid chromatography. *Planta* **96**, 271–280.

Little, C. H. A., Heald, J. R. and Browning, G. (1978). Identification and measurement of indoleacetic acid and abscisic acids in the cambial region of *Picea sitchensis* (Bong.) Carr. by combined gas chromatography–mass spectrometry. *Planta* **139**, 133–138.

Liu, S-j. and Tillberg, E. (1983). Three-phase extraction and partitioning with the aid of dialysis—a new method for purification of indolyl-3-acetic and abscisic acids in plant materials. *Physiol. Plant.* **57**, 441–447.

Loveys, B. R. (1977). The intracellular location of abscisic acid in stressed and non-stressed leaf tissue. *Physiol. Plant* **40**, 6–10.

Loveys, B. R. and Milborrow, B. V. (1981). Isolation and characterisation of 1′-*O*-Abscisic acid-β-D-glucopyranoside from vegetative tomato tissue. *Aust. J. Plant Physiol.* **8**, 571–589.

Mapelli, S. and Rocchi, P. (1983). Separation and quantification of abscisic acid and its metabolites by high-performance liquid chromatography. *Ann. Bot.* **52**, 407–409.

Martin, G. C., Dennis Jnr., F. G., Gaskin, P. and MacMillan, J. (1977). Identification of gibberellins A_{17}, A_{25}, A_{45}, abscisic acid, phaseic acid and dihydrophaseic acid in seeds of *Pyrus communis*. *Phytochemistry* **16**, 605–607.

Mertens, R., Stüning, M. and Weiler, E. W. (1982). Metabolism of tritiated enantiomers of abscisic acid prepared by immunoaffinity chromatography. *Naturwissenschaften* **69**, S, 595.

Mertens, R., Deus-Neumann, B. and Weiler, E. W. (1983). Monoclonal antibodies for the detection and quantitation of the endogenous plant growth regulator, abscisic acid. *FEBS Lett.* **160**, 269–272.

Milborrow, B. V. (1967). The identification of (+)-abscisin II [(±)-dormin] in plants and measurement of its concentrations. *Planta* **76**, 93–113.

Milborrow, B. V. (1972). The biosynthesis and degradation of abscisic acid. *In* "Plant Growth Substances 1970" (D. J. Carr, ed.), pp. 281–290. Springer-Verlag, Berlin.

Milborrow, B. V. (1978). Abscisic acid. *In* "Phytohormones and Related Compounds—A Comprehensive Treatise". Vol. 1. (D. S. Letham, P. B. Goodwin and T. V. J. Higgins, eds), pp. 295–347. Elsevier/North Holland, Amsterdam.

Milborrow, B. V. (1983a). Biosynthesis of abscisic acid and related compounds. *In* "Biosynthesis of Isoprenoid Compounds". Vol. 2. (J. W. Porter and S. L. Spurgeon, eds), pp. 413–436. Academic Press.

Milborrow, B. V. (1983b). Pathways to and from abscisic acid. *In* "Abscisic Acid" (F. T. Addicott, ed.) pp. 79–111. Praeger, New York.

Milborrow, B. V. (1983c). The reduction of (±)-[2-^{14}C] abscisic acid to the 1′,4′-*trans* diol by pea seedlings and the formation of 4′-desoxy ABA as an artefact. *J. Exp. Bot.* **34**, 303–308.

Milborrow, B. V. and Mallaby, R. (1975). Occurrence of methyl (+)-abscisate as an artefact of extraction. *J. Exp. Bot.* **26**, 741–748.

Milborrow, B. V. and Noddle, R. C. (1970). Conversion of 5-(1,2-epoxy-2,6,6-trimethylcyclohexyl)-3-methylpenta-*cis*-2-*trans*-4-dienoic acid into abscisic acid in plants. *Biochem. J.* **119**, 727–734.

Milborrow, B. V. and Vaughan, G. T. (1982). Characterisation of dihydrophaseic acid 4′-*O*-β-D-glycopyranoside as a major metabolite of abscisic acid. *Aust. J. Plant Physiol.* **9**, 361–372.

Millard, B. J. (1979). "Quantitative Mass Spectrometry." Heyden, London.

Most, B. H. (1971). Abscisic acid in immature apical tissue of sugar cane and in leaves subjected to drought. *Planta* **101**, 67–75.

Mousale, D. M. A. (1981). Reversed-phase ion-pair high performance liquid chromatography of the plant hormones indolyl-3-acetic and abscisic acids. *J. Chromatogr.* **209**, 489–493.

Mousdale, D. M. A. and Knee, M. (1979). Poly-*N*-vinylpyrrolidone column chromatography of plant hormones with methanol as eluent. *J. Chromatogr.* **177**, 398–400.

Neill, S. J. and Horgan, R. (1983). Incorporation of α-ionylidene ethanol and α-ionylidene acetic acid into ABA by *Cercospora rosicola. Phytochemistry* **22**, 2469–2472.

Neill, S. J., Horgan, R. and Heald, J. K. (1983). Determination of the levels of abscisic acid-glucose ester in plants. *Planta* **157**, 371–375.

Neill, S. J. Horgan, R. and Walton, D. C. (1984). Biosynthesis of abscisic acid. *In* "The Biosynthesis and Metabolism of Plant Hormones." Society for Experimental Biology Seminar Series 23 (A. Crozier and J. R. Hillman, eds), pp. 43–70. Cambridge University Press, Cambridge.

Netting, A. G., Milborrow, B. V. and Duffield, A. M. (1982). Determination of abscisic acid in *Eucalyptus haemastoma* leaves using gas chromatography–mass spectrometry and deuterated internal standards. *Phytochemistry* **21**, 385–389.

Norman, S. M., Maier, V. P. and Echols, L. C. (1982). Reversed-phase clean-up and high performance liquid chromatographic analysis of abscisic acid in *Cercospora rosicola* liquid culture media. *J. Liq. Chromatogr.* **5**, 81–91.

Pierce, M. and Raschke, K. (1981). Synthesis and metabolism of abscisic acid in detached leaves of *Phaseolus vulgaris* L. after loss and recovery of turgor. *Planta* **153**, 156–165.

Quarrie, S. A. (1978). A rapid and sensitive assay for abscisic acid using ethyl abscisate as an internal standard. *Anal. Biochem.* **87**, 148–156.

Rajasekaran, K., Vine, J. and Mullins, M. G. (1982). Dormancy in somatic embryos and seeds of *Vitis*: changes in endogenous abscisic acid during embryogeny and germination. *Planta* **154**, 139–144.

Rivier, L., Milon, H. and Pilet, P.-E. (1977). Gas chromatography–mass spectrometry determinations of abscisic acid levels in the cap and apex of maize roots. *Planta* **134**, 23–27.

Rivier, L. and Saugy, M. (1986). Chemical ionization mass spectrometry of IAA and ABA: evaluation of negative ion detection and quantification of ABA in growing maize roots. *J. Plant Growth Regul.* **5**, 1–16.

Roberts, D. L., Heckman, R. A., Hege, B. P. and Bellin, S. A. (1968). Synthesis of (*RS*)-abscisic acid. *J. Org. Chem.* **33**, 3566–3569.

Rosher, P. H., Jones, H. G. and Hedden, P. (1985). Validation of a radioimmunoassay for (+)-abscisic acid in extracts of apple and sweet-pepper tissue using high-pressure liquid chromatography and combined gas chromatography–mass spectrometry. *Planta* **165**, 91–99.

Saunders, P. F. (1978). The identification and quantitative analysis of abscisic acid in plant extracts. In "Isolation of Plant Growth Substances" Society for Experimental Biology Seminar Series 4 (J. R. Hillman, ed.), pp. 115–134. Cambridge University Press, Cambridge.

Schlenk, M. and Gellerman, J. L. (1960). Esterification of fatty acids with diazomethane on a small scale. *Anal. Chem.* **32**, 1412–1414.

Seeley, S. D. and Powell, L. E. (1970). Electron capture–gas chromatography for sensitive assay of abscisic acid. *Anal. Biochem.* **35**, 530–533.

Setter, T. L., Brenner, M. L. and Brun, W. A. (1981). Abscisic acid translocation and metabolism in soybean following depodding and petiole girdling treatments. *Plant Physiol.* **67**, 774–779.

Sivakumaran, S., Horgan, R., Heald, J. and Hall, M. A. (1980). Effect of water-stress on metabolism of abscisic acid in *Populus robusta* × schneid and *Euphorbia lathyrus* L. *Plant, Cell and Environment* **3**, 163–173.

Sondheimer, E. and Tinelli, E. T. (1971). Improved synthesis of (*RS*)-2-^{14}C- abscisic acid. *Phytochemistry* **10**, 1663–1664.

Sondheimer, E., Galson, E. C., Chang, Y. P. and Walton, D. C. (1971). Asymmetry, its importance to the action and metabolism of abscisic acid. *Science* **174**, 829–831.

Sotta, B,. Sossountzov, L., Maldiney, R., Sabbagh, I., Tachon, P. and Miginiac, E. (1985). Abscisic acid localisation by light microscopic immunohistochemistry in *Chenopodium polyspermum* L. *J. Histochemistry and Cytochemistry* **33**, 201–208.

Steen, I. and Eliasson, L. (1969). Separation of growth regulators from *Picea abies* Karst. on Sephadex LH-20. *J. Chromatogr.* **43**, 558–560.

Sweetser, P. B. and Vatvars, A. (1976). High performance liquid chromatographic analysis of abscisic acid in plant extracts. *Anal. Biochem.* **71**, 68–78.

Taylor, H. F. and Burden, R. S. (1970). Identification of plant growth inhibitors produced by photolysis of violaxanthin. *Phytochemistry* **9**, 2217–2223.

Taylor, H. F. and Burden, R. S. (1972). Xanthoxin, a recently discovered plant growth inhibitor. *Proc. R. Soc. Lond.* B **180**, 317–346.

Taylor, H. F. and Burden, R. S. (1973). Preparation and metabolism of 2-[^{14}C]-*cis,trans*-xanthoxin. *J. Exp. Bot.* **24**, 873–880.

Taylor, I. B. and Rossall, S. (1982). The genetic relationship between the tomato mutants *flacca* and *lateral suppressor,* with reference to abscisic acid accumulation. *Planta* **154**, 1–5.

Tietz, D., Dörffling, K., Wöhrle, D., Erxleben, I., and Liemann, F. (1979). Identification by combined gas chromatography–mass spectrometry of phaseic acid and dihydrophaseic acid and characterization of further abscisic acid metabolites in pea seedlings. *Planta* **147**, 168–173.

Vaughan, G. T. and Milborrow, B. V. (1984). The resolution by HPLC of *RS*-[2-^{14}C] Me 1′,4′-*cis*-diol of abscisic acid and the metabolism of (−)-*R*- and (+)-*S*-abscisic acid. *J. Exp. Bot.* **35**, 110–120.

Velaso, J., Ram Chandra, G. and Mandava, N. (1978). Derivatisation of abscisic acid as the p-nitrobenzyl ester. *J. Agric. Food Chem.* **26**, 1061–1064.

Walton, D. C. (1980). Biochemistry and physiology of abscisic acid. *Annu. Rev. Plant Physiol.* **31**, 453–489.

Walton, D. C. and Sondheimer, E. (1972). Metabolism of 2-^{14}C-(±)-abscisic acid in excised bean axes. *Plant Physiol.* **49**, 285–289.

Walton, D. C., Wellner, R. and Horgan, R. (1977). Synthesis of tritiated ABA of high specific activity. *Phytochemistry* **16**, 1059–1061.

Walton, D. C., Dashek, W. and Galson, E. (1979). A radioimmunoassay for abscisic acid. *Planta* **146**, 139–145.

Weiler, E. W. (1979). Radioimmunoassay for the determination of free and conjugated abscisic acid. *Planta* **144**, 255–263.

Weiler, E. W. (1980). Radioimmunoassays for the differential and direct analysis of free and conjugated abscisic acid in plant extracts. *Planta* **148**, 262–272.

Weiler, E. W. (1982). Enzyme-immunoassay for *cis*-(+)-abscisic acid. *Physiol. Plant.* **54**, 510–514.

Wheaton, T. A. and Bauscher, M. G. (1977). Separation and identification of endogenous growth regulators in citrus. *Proc. Int. Soc. Citricult.* **2**, 673–676.

Whenham, R. J. and Fraser, R. S. S. (1981). A rapid and simple radioassay for abscisic acid using ^{14}C-diazomethane. *J. Exp. Bot.* **32**, 1223–1230.

Yokota, T., Murofushi, N. and Takahashi, N. (1980). Extraction, purification and identification. In "Hormonal Regulation of Development. I. Molecular Aspects of Plant Hormones". Encyclopaedia of Plant Physiology, New Series, Vol. 9 (J. Macmillan, ed.), pp. 113–202. Springer-Verlag, Berlin.

Zeevaart, J. A. D. (1974). Levels of (+)-abscisic acid and xanthoxin in spinach under different environmental conditions. *Plant Physiol.* **53**, 644–648.

Zeevaart, J. A. D. (1983). Metabolism of abscisic acid and its regulation in *Xanthium* leaves during and after water stress. *Plant Physiol.* **71**, 477–481.

Zeevaart, J. A. D. and Milborrow, B. V. (1976). Metabolism of abscisic acid and the occurrence of epi-dihydrophaseic acid in *Phaseolus vulgaris. Phytochemistry* **15**, 493–500.

Index

The following abbreviations are used in this index:

ABA	Abscisic acid
GA	Gibberellin
GC	Gas chromatography
HPLC	High performance liquid chromatography
IAA	Indole-3-acetic acid
MIM	Multiple ion monitoring
MS	Mass spectrometry
NMR	Nuclear magnetic resonance
SIM	Single ion monitoring
TLC	Thin layer chromatography

YORK COLLEGE OF PENNSYLVANIA 17403

DISCARDED
LIBRARY